The Woodwright's Guide

Also by Roy Underhill

The Woodwright's Shop (1981)

The Woodwright's Companion (1983)

The Woodwright's Workbook (1986)

The Woodwright's Eclectic Workshop (1991)

The Woodwright's Apprentice (1996)

Khrushchev's Shoe (2000)

The Woodwright's Guide

Working Wood with Wedge and Edge

Roy Underhill

Illustrations by Eleanor Underhill

The University of
North Carolina Press
Chapel Hill

Designed by Kimberly Bryant
Set in The Serif and The Sans by Rebecca Evans
Manufactured in the United States of America

The paper in this book meets the guidelines for
permanence and durability of the Committee on
Production Guidelines for Book Longevity of the
Council on Library Resources.

The University of North Carolina Press has been
a member of the Green Press Initiative since 2003.

LIBRARY OF CONGRESS CATALOGING-
IN-PUBLICATION DATA

Underhill, Roy.
The woodwright's guide : working wood with
wedge and edge / Roy Underhill ; illustrations
by Eleanor Underhill.
    p. cm.
Includes bibliographical references and index.
ISBN 978-0-8078-3245-5 (cloth : alk. paper)
ISBN 978-0-8078-5914-8 (pbk. : alk. paper)
1. Woodworking tools. 2. Wedges I. Title.
TT186.U64 2008
684′.082—dc22   2008024431

cloth   12 11 10 09 08 5 4 3 2 1
paper   12 11 10 09 08 5 4 3 2 1

University of North Carolina Press books may be
purchased at a discount for educational, business,
or sales promotional use. For information, please
visit www.uncpress.unc.edu or write to UNC Press,
attention: Sales Department, 116 South Boundary
Street, Chapel Hill, NC 27514-3808.

for the Galoots

Contents

The Woodwright's Guide

Introduction

My favorite comic book—snitched from Mark Olshaker's stash and read by flashlight under my blanket at summer camp—was *Atomic Knights!* In its post-Armageddon world, underground survivors dared not risk walking Earth's radioactive, rubble-strewn surface. Not, that is, until scientists discovered that old suits of armor would somehow protect the wearers from the still-lethal atomic radiation. Thus, our heroes, the Atomic Knights, ventured forth from the shelters to battle re-emergent evil, armed with swords, safely clad in suits of ancient iron.

Since then, I've learned to recognize the archetypal myth of redemption by ancestral spirits. In myth, it's always a sword, never a chisel. Still, it's the wood-working blade that got us here today. The ways in which we work make an ever greater difference to this ever smaller planet. A world that doesn't end with a communist bang can still go down with a consumerist whimper.

The working blade cuts deep into our history. An old axe or chisel likely contains iron that has been recycled since Roman times. The iron wedge and steel edge and the grain of wood are still with us. We still use a wedge to split the wood, exploiting the planes of weakness in the grain—paradoxically capturing its strength. We still use an edge to shear the wood, exposing the beauty of the grain, shaping it to our desire. Wedge and edge—obvious at times, sometimes working unseen and side by side, just as they have for thousands of years.

This book's journey begins in the forest and passes through each woodworking trade as its moves farther from the tree. We go from forest to furniture, from green to dry, from risk to certainty, nature to culture, outdoor to indoor, multiple hernias to carpal tunnel syndrome. I describe the tools and connections as they arise in each working environment. Thus, once the technique of drawboring a mortise and tenon joint is introduced in the carpenter's work, it reappears as a known quantity in joinery and cabinetmaking.

I revisit here many of the techniques that I discussed in earlier books. I hope I have explained matters with greater clarity and presented fewer idiosyncratic methods. Over the years, I've had the opportunity to watch some true masters at work. To paraphrase Isaac Newton, "I have looked over the shoulders of giants." I'll try to get out of the way as much as I can and let you see for yourself.

1 Faller

*We experimented, as young boys will, and we felled one large
hickory with the saw instead of the axe, and barely escaped with
our lives when it suddenly split near the bark, and the butt shot
out between us. I preferred buckeye and sycamore for my own axe;
they were of no use when felled, but they chopped delightfully.*
    —William Dean Howells, My Year in a Log Cabin, 1893

It's just a piece of steel on the end of a stick, but let's see what your axe has to say. Hold the end of the helve in one hand and, with the other hand held close to the axe head, carry it up over your shoulder. Start the swing, sliding your top hand down to meet the other. The springy hickory helve stores part of the energy from the start of the swing and then releases it near the end of the swing. Just as the end of a whip breaks the sound barrier, the axe head accelerates over the arc of the swing to the instant of impact with the wood. Right now the wood still belongs to the tree. Here's where it all begins.

You'll first cut the notch on the side of the tree facing the direction in which you want it to fall—but don't cut yet. This is dangerous. First, see if the tree is leaning or heavier on one side. If the tree really leans, consider dropping it at right angles to the lean. Dropping in the same direction as a severe lean can cause the tree to split and go over before you're ready. Guy ropes placed high in the tree can encourage it to fall where you want but also add the dangers of climbing and of ropes under high tension.

Find a clear path for the tree to hinge to the ground. Branches, dead or living, can break free and spear you. If the tree you're felling hangs up in another tree, you may have to drop them both—never a safe or clean process. Consider the landing area. A hump or hollow can crack the log when it lands. A tree landing on an upward slope is liable to jump downhill. Wherever it lands, figure out how you're going to get the logs out.

Take nothing for granted. Get everyone and everything out of the way. I've dropped many trees, but the last one landed right on my mailbox, demolishing it. Amusing to the neighbors, but not my intention. You can determine where the top of the tree should end up by stepping back from the tree at right angles to the direction of the fall. Hold your axe handle in your extended arm and sight the top and bottom of the tree relative to it. Now, keeping the end of the handle aligned with the base of the tree, pivot the axe handle in the direction of the fall. It projects a scale model of the falling tree onto the landscape.

Now we're ready to chop for real. Just for now, make the first swing straight in at the tree. The edge of the axe makes contact from the center outward, easing the shock of impact. The middle of the curve of the sharp bit of the axe intersects the bundled fibers of the tree at right angles to their length. The curve of the bit also makes all but the center of the bit cut with a shear and allows greater penetration for the leading edge. The sharp blade has struck the tree with tremendous force—but without much result.

The steel edge severed the fibers, pushing them aside, but the pressure of the compressed wood on either side of the cut quickly stopped the progress of the blade. With this straight-in blow, the wedging action halted the edge action. Not much happened, but the next strike will not be like the first.

The next stroke comes in higher on the tree than the first and at a downward-sloping angle. Gravity adds its force to the swing, but, more important, now the wedge action of the axe is your helper. The edge severs the fibers, and as the compression builds up on the cheeks of the blade, the wood splits apart—the chip moving out on the face of least resistance. The split continues downward until it reaches the first cut. Now, the chip of wood moves outward and relieves the pressure on the cheeks. The edge continues far deeper into the wood before its momentum is spent. The converging blows and the splitting out of chips between them fell the tree. Little strokes fell great oaks.

When you need a break, stop to study one of the chips. You'll see that the area showing split marks is far greater than that cut by the edge. You'll also feel the water in the tree. A living tree is about half water by weight. This water fills the cells, swelling them, making the wood softer and easier to cut and split than when it dries out. All woodworking takes place on a continuum of wood moving from wet to dry. Right now, we want the wood green and easy to cut and split. Later, we'll go to great lengths to prevent the wood from splitting. We'll move from the risk of the axe to the certainty of the plane.

Back to work. The edge of the axe cuts in, and the cheeks split chips away, creating an ever-deepening notch. The notch is relatively flat bottomed, with a top slope of about 45 degrees. The notch widens as it deepens, until it reaches just beyond halfway through the tree. The tree is now unsupported on that side and begins the slightest lean in that direction.

Now to the back cut. It's good to have someone with you. First, for safety. Second, your companion can pull the other end of the saw for the cut that drops the tree. You can drop the tree with another axe notch, but the saw cuts faster and leaves less torn fiber in the hinge.

With one of you on each end of the saw, set the teeth a few inches above the level of the axe cut and give it a start. You may find it easier to support the middle of the saw with your hand or toe until you get a kerf going. Say "to me," or something of the sort, so your comrade will know that you're about to pull.

You're always advised to pull a crosscut saw, never push it. Pushing a long "misery whip" is like pushing a rope, causing it to bend in the kerf and drag. Still, you should feed the saw and your arms back to your partner on the return stroke. Don't bear hard into the cut, but rather pull the saw across the surface and "let the saw do the work." Right.

No matter how well the saw is cutting, the dreaded pinch may await you as you saw through the still-standing tree. A slight ill breeze and the tree can lean back the opposite way from where you intended. You can drive an iron wedge into the kerf behind the saw, but this is a better preventive than a cure. Until the breeze turns in your favor, all the wedges in the world aren't going to tip that tree back over. You simply wait.

The idea of the front notch and back cut is to hinge the tree down—the intact wood between the two cuts holds together until it finally snaps like a popsicle stick. As you get closer to the middle, the tree will start to give signs of what it's going to do—nodding in the direction where it wants to go.

Get ready now. Both of you need a clear path to get away to the sides, and never to the back, where the butt might shoot. Cut as fast as you can toward the end. You want this hinge to be just thick enough to direct the fall but not slow it. You need the unimpeded velocity of the fall to drive the top of the tree through the surrounding branches.

It's going over now. Branches cracking, leaves falling, the tree twists and quivers on its way down. It lands and rolls. The surrounding trees sway. The final few leaves are still drifting down as you walk back, glancing up at the canopy for still-hanging snags, blinking at the sunlight you've let into the forest, not quite ready to look squarely at your fallen tree.

Helves

How's your axe? If the head has been creeping up (about to fly off the handle), you may just need to drive the wedges a bit more—or it may be time for a new helve. If enough meat of the hickory helve remains, you can drive the helve farther through the head to freshen the wedged end. You might need to shave away some of the shoulder below the head, paying close attention to the hang of the axe. The bit of the axe should be in line with the helve and slightly "closed"—tilted toward the tail end of the helve to keep the bit at right angles to the arc of the swing.

New helve or old, drive it through the axe head using inertia as your anvil. Hold the axe in one hand with the head hanging down and smack the butt end of the helve with a mallet. Inertial mass will keep the heavy axe head still as the helve drives in.

If you are driving an existing helve deeper into a head, you can remove or reset the old wedges by carefully sawing away any protruding handle. This may expose enough of a wedge so that you can tap it from side to side and work it free, or drive it deeper. If you need a larger iron wedge, a smith can easily make one or you can likely find one at a good hardware store.

You also have to remove the iron wedge if you need to replace the central wooden wedge. The wooden wedge must be hard enough to resist crushing, but not so hard that it won't conform and grip in the slot sawn in the end of the helve. Glue on the wedge helps it grip, and the steel wedge driven in diagonally across it locks it in place. A softer wooden wedge may take the steel wedge with less chance of splitting, but I still use hickory, split from the same billet used to make the helve.

Just as the wood splits in one direction and not in the other, it also swells and shrinks unevenly. Like age on a man, water makes wood softer, heavier, and fatter—but not taller. Tightly fit an axe head with a handle made out of unseasoned wood and check it six months later. It will still measure the same length but will knock around in the axe head like the clapper in a bell. Ideally, then, you'll keep a small stack of hickory, ash, locust, or maple billets seasoning in your loft—drying for years before you need them for handles and helves.

Once dry, wood remains hygroscopic, taking in or releasing water in balance with its environment. Henry David Thoreau may have stuck the head of his borrowed axe in Walden Pond to swell and tighten the helve—but don't you do it. Soaking a hickory axe helve in water swells the wood, making it absolutely tight in the head—for a while. The water expands the wood so much that it is crushed against the unyielding walls of the axe eye. Upon drying, however, the fibers shrink back smaller than before, and Thoreau's axe head goes flying off—Oops! Sorry, Mrs. Emerson.

Drive the helve into the eye as inertia holds the head.

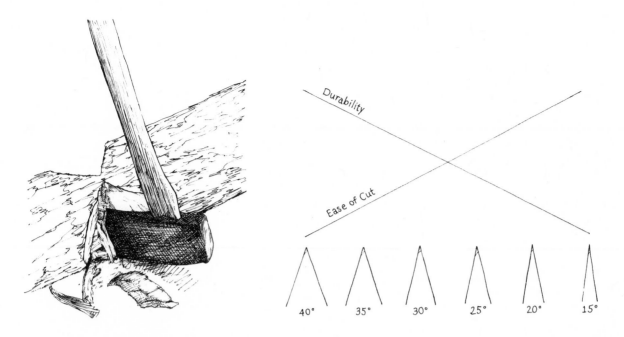

Buck and Sharpen

Back to the tree. Remove the branches by swinging up from below at their base, with perhaps a finishing blow into the crotch. You may want to leave a branch as a lever to help you turn the log, or to keep it off the ground. Nothing is stable yet, so stay alert.

Bucking with an axe is fine for smaller trees. Stand on top and swing down below and between your feet, opening a notch almost as wide as the log is thick. Chop halfway through from one side and halfway from the other.

The work is slowing down. Is it you, or does the axe need sharpening? Cutting wood means sharpening tools, and sharpening means compromise, or if you prefer, creating the optimal balance of wedge and edge.

With an axe, wedge and edge work together, each allowing the other to do its job. The more acute the edge, the more easily and deeply it can cut the fibers before the splitting action begins. If the edge is too obtuse, too wedgelike, the resistance of the compressed fibers will stop the penetration of the edge too quickly. But if the edge is excessively acute, the bit may sink in deeply without any wedging action, sticking in the wood and never popping out a chip.

In edge tools, another factor comes into play—durability, or how long the edge will last. Any cutting edge is the intersection of two lines, both in theory and reality. With increasing acuity, the line of cutting ease goes up as the line of durability goes down. The durability line slides to the right with better steel, to the left with harder wood. Optimum sharpening lies at the intersection of these two lines.

After repeated use and sharpening, an axe tends to become more wedge-like, more obtuse. So, it's not just the edge that needs touching up; the cheeks need to be brought in as well. This means you have to remove a lot of steel, but axes, along with hatchets, saws, and auger bits, are tempered soft enough to sharpen easily with a file.

Draw the file along the bit.

You'll need both hands on your file, so sit on a log and slip the axe handle under one leg so that the head presses firmly on your opposing knee. Then, push or draw the fine, flat file along the full length of the blade. Take care—you can get badly cut.

The file cuts in only one direction, so lift it off the steel on the return stroke. On most other tools, the sharpening bevels, the faces that intersect to form the edge, are flat or faceted. On axes and knives, leave these bevels gently rounded. When cutting harder woods, you want more rounding and a more obtuse edge, upward of 30 degrees. On softer stuff, you can go down below 25 degrees. You'll soon learn what your steel can take.

On the cheeks of an old axe, up away from the edge, you may not be filing steel at all—it may be wrought iron. Old tools are mostly forged from tough, but too-soft-to-hold-an-edge wrought iron, with only the bit made of more expensive, more fragile carbon steel. This combination of steel and iron gives old tools their uncompromised toughness, and (in both senses of the word) it gives them their edge.

On an axe head, the layer of steel goes right down the center of the business end. On old chisels, plane irons, drawknives and the like, you'll see the layer of tool steel on the flat side of the iron body. Unlike axes, these other tools are not exposed to violent shock and frozen knots, so their steel can be hardened beyond the ability of a file to cut them. Axes, to endure, need their softer temper.

Crosscut Saw

The axe is sharp again, but now let's buck the tree into logs with the big crosscut saw. Whereas the axe is a versatile individualist, the saw is a team of specialists. One hundred teeth work in teams to cut as many tiny chips as they can carry away with every stroke. Just as the axe made an opening wider than itself by cutting on either side of a cross-grain notch, so does the saw, slicing across the grain on alternate sides of the kerf. Unlike the symmetrically edged felling axe, saw teeth are flat on the outside and beveled to the inside of the cut. This asymmetry forces the tiny chip to the inside of the kerf where it gets carried out in the spaces between the teeth.

This log won't pinch.

Sharpening Big Crosscut Saws

Raker-toothed crosscut saws operate on the same principle you find in planes that work across the grain. They first slice across the grain with knifelike blades and then shave out the wood in between. Learn to sharpen them and they won't be such "misery whips." When all is right, the saw works easily and pulls out long strings of wood, severed by the slicers and planed free by the rakers.

The crosscut saw can make its own filing vise out in the woods. Saw a kerf in a log just deep enough so that you can set the saw in it with the teeth sticking up. You could also lay the saw on a stump and file the bit that hangs over, or you can wait until you get back and sharpen the saw while holding it in a proper clamp. In any case, the steps are jointing, fitting the rakers, sharpening the cutters, and finally, setting.

Jointing brings all the slicing teeth to the same height. Lightly pull a file down the tops of the teeth, taking care to hold the file exactly square to the sides of the blade. Stop when you have brightened the tips of all the slicing teeth.

The rakers need to sit slightly farther back than the other teeth, about 1/100 inch for hard wood and 3/100 inch for soft. Combination sharpening tools have a bridge that rides on the tips of the teeth and an adjustable window to drop over the rakers. You can set the depth of the window with a feeler gauge (or with the thickness of a matchbook cover) to guide your file precisely, topping the rakers at the proper depth.

The cutters and the rakers are now at their proper and consistent lengths, but flat tipped. Sharpen the rakers by smoothing their vertical face square across and then filing the back slope. The flat should almost disappear, but not quite. Stop when one more stroke of the file would take away that last bit of the flat tip.

Shape the profile of the slicing teeth to a slightly rounded 60-degree point, with the bevels from between 30 degrees for soft wood and 45 degrees for hard. Again, file until the flats from jointing almost disappear.

Set the teeth by either bending the tips away from the bevel side with a slotted saw wrest or hammer-setting them on a stake anvil or against an axe head or sledgehammer held in one hand. All you need is something with enough inertial mass to stand up to the eight-ounce setting-hammer blows delivered with your other hand. Mount the saw upright in a vise and position the anvil 1/4 inch below the tip of a tooth. Strike the base of the tooth bevel, bending it over the anvil. A set of 1/100 inch is good for hard wood and up to 3/100 for soft wood. Measure the set of the teeth with a saw-setting spider, a three- or four-legged rocker gauge. If the spider were a table, one leg would be short by the amount of set you desire. Move on down the saw, setting and checking every other tooth to one side and then completing the other side.

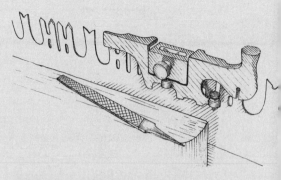

File the rakers using the guide on a combination tool. This one also has a saw wrest, a hammer, and an adjustable three-legged spider (lower right).

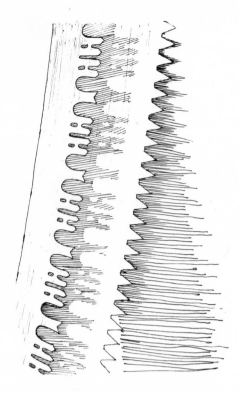

A perforated lance-toothed saw casts its shadow on a peg-toothed crosscut.

In simple peg-toothed saws, the teeth are having to do two jobs. They have to slice across the fibers, and they also have to break the tiny chip, force it to the middle, and carry it out of the cut. If we made them sharper, more knifelike, they would slice better—but then they wouldn't break the chips and carry them out as well. We can only make the saw teeth sharper if we add another type of tooth to the team, one that can both break the chip and carry it away. This is the raker tooth. It rides a little bit back from the points of the cutters, allowing them to do their job before routing out the wood left in the middle of the cut.

None of these teeth will get very far if the saw doesn't have enough set—the alternating outward bending of the teeth that allows the saw to make a kerf wider than its thickness. If we take the practical maximum set of a saw tooth as one-third the thickness of the blade, then the kerf could be two-thirds wider than the blade is thick. This keeps the blade from binding, but it also means you're cutting two-thirds more wood. You want enough set to keep the saw cutting freely, but no more than that.

Olive oil on the saw can keep pine sap from gumming up the works, but neither set nor soap can keep a saw moving in a pinched kerf. If you have arranged the landing spot so that the log is supported in the middle, the kerf will open as you saw. But when a log sits so that it is supported more at the ends, you'll get halfway through and gradually find your saw seized by several tons of pressure.

Wedges driven into the kerf above the saw may keep it open, but wedges won't open a pinch in a heavy log. Try to roll the log over and continue the cut from the side. Of course you can't roll a log with a saw pinched in it, so all you can do is try to lever the log up and jam a support beneath it.

If none of this works (and assuming you can get the saw free from the kerf), you can make a second saw kerf parallel to and an inch or so over from the first. When this kerf starts to pinch, pull the saw out and, standing on top of the log, swing down with the poll of your axe and pop out the intervening wood and continue sawing. This is a neat trick, but it hardly redeems the nuisance.

LEFT: *A cant hook helps you roll the logs about . . .*

RIGHT: *. . . but a timber carrier lets you share the fun.*

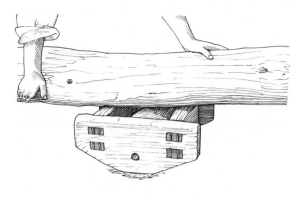

The "drug" — a cylindrical roller mounted in a wooden frame.

Moving

Felling the tree is often easier than getting it out of the woods. Because they roll easily, short lengths of big trees are easiest to move. If you need to turn the direction of a rolling log, roll it up onto something like a limb stub to elevate it at its balance point. If it's balanced on a pivot point, you can probably turn it.

A timber is sometimes called a cant; thus you roll it with a lever called a cant hook. Two cant hooks opposed can help two people carry a timber out of the woods, but the pivot connection on a log carrier makes negotiating between trees far easier. If you have several timber carriers and several pairs of people working, remember that the closer to the middle a carrier is set, the more load that team will carry.

If you can get one end up on a skid so it won't dig in, horsepower can pull the log back to the road. Rollers, either short logs cut on site or a contrivance mounted in a frame, can also help for short distances. What you really need, though, is a pair of wheels — big ones mounted on an axle that can straddle the log at its balance point, lift it, and roll it away. The straddling and rolling are obvious, but the lifting — not so much. Although some carts use a capstan or even a big iron screw to hoist the log, the simplest and most common solution employs the axle as a fulcrum and the long tongue of the cart as a lever.

Some timber carts have their axle in an arch. Lifting the tongue drops the top of the arched axle against the log so that they can be chained close together. Pulling down on the long tongue raises the arch and lifts the log with considerable mechanical advantage.

Lacking an arched axle, an extended tongue will also do the trick. Like a seesaw offset to balance unequal loads, the tongue mounts on the axle with about a fifth of its length extending to the back. Again, you straddle the log with the axle over the balance point. Lift the tongue so that the short extension drops to touch the log. Chain the log to this end and have everyone push down on the tongue — lifting the log. The log may be lifted, but it is not yet safely suspended. Should one of you step back to admire the work before you chain the log to the long end of the tongue, the others will be catapulted.

2 Cleaver and Countryman

When frost will not suffer to dyke and to hedge,
then get thee a heat with thy beetle and wedge:
A short saw, and long saw, to cut a-two logs,
an axe, and an adze, to make trough for thy hogs;
A grindstone, a whetstone, a hatchet and bill,
with hamer, and english naile, sorted with skil;
A frower of iron, for cleaving of lath,
with roll for a sawpit, good husbandrie hath.
    —after Thomas Tusser,
    Five Hundred Points of Husbandry, 1557

Exploit the weakness of the grain when you work wood . . .

. . . and its strength when you use it.

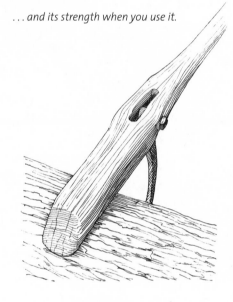

Clearing land, splitting fence rails, shaving shovel handles, bending oxbows—there are few aspects of woodworking unfamiliar to the countryman. He fells the hickory, splits it and shaves it to make a chair, strips its bark to weave the chair bottom, and sits in it by a fire fueled by the limbs. The faller may never hold a plane; the cabinetmaker may not know where wood comes from—the countryman sees it all.

It begins in the grain of the wood. We learn to exploit the weakness of the grain when we work wood, and to exploit the strength of the wood when we use it. Wagon spokes, chair rungs, hayforks—anything where strength is essential—all are riven from the log. Sawing would take more energy and investment. Worse, it ignores the path of the grain, too often cutting across it. Sawn stock—weaker and costlier. Riven stock—cheaper and stronger.

Splitting wood makes the tree a partner with you. I found the best expression of this partnership in a cant hook I bought at a farm auction. The shaft of the cant hook was shaved from a riven billet of white oak, exploiting the weakness of the wood. The split followed the grain, ensuring that all the strength of the wood was captured within the shaft.

The riving also revealed a knot near one end of the stave. A knot (a branch overgrown by the tree) is usually considered a weak point in a piece of wood—but that is true only for sawn timber. As the tree grows outward, it swallows the branch with the grain flowing around it like a river around a rock. The straight path of a saw knows nothing of this and cuts through the flow of the grain. Splitting, however, followed the grain and left a bulbous section around the knot. The maker of the cant hook followed the grain as well, putting the hole for the iron hook right in the middle of the flow.

In riving, you usually split a piece in half, then split that piece in half, and so forth. Halving each piece helps the split run straight—as long as the grain is straight. Often it's not, but because wood and bark are made in the same layer in the tree, you can usually tell the book by the cover. If the bark spirals, so too will the wood within.

Oaks, tulip poplar, hickories, ash, chestnut, pine, cypress, walnut, cedar, and basswood are among the most cooperative splitters. All of them split more easily when the wood is fresh. Of course, if it's an elm or a gum log you're looking at, forget it.

Splitting is not without hazards to health and hardware. Considering all the times I've hit steel wedges with a steel sledge, I'm going to hell for sure. The steel sledge on the steel wedge gradually work-hardens the head of the wedge so that one fatal strike can send shrapnel flying. A worse sin is using the head of an axe to drive a serious wedge. This quickly bows out the thin eye of the axe and destroys it. Using the axe head as a wedge and driving it with a heavy sledge has the same effect. If your axe is the only wedge you have with you, a good roundhouse swing at the end grain of a short log might split it or might start a split that you can advance with another well-placed strike with the axe or more proper splitting tools.

Oak billets for the shingle maker.

LEFT: *The beveled striking head keeps the glut from splintering.*

RIGHT: *A hickory root maul drives dogwood wedges to split a red oak log.*

Beetle and Wedge

You'll need an axe or a steel wedge to start a split, but once it's begun, you can continue the split with gluts — wooden wedges chopped from any hard, tough wood. Gluts can't have quite as much taper as a steel wedge, but they can be made much longer. You can use a two-foot-long glut to great advantage on a big log.

The poet says, "Get thee a heat with thy beetle and wedge." "Beetle" is the Old English word for a mallet to pound anything from pavement to laundry. You may work up a sweat, but the beetle takes a beating. Tough elm makes a good beetle head, and binding it with iron rings will keep the edges from chunking off.

This kind of mallet with a separate head and handle makes for an end grain striking face, but what of the one-piece maul where the striking face is side grain? One-piece mauls for driving wedges are often called root mauls, because you make the striking head from the underground part of the tree. All those roots radiating from the base of the stem peg the striking head together and buttress the faces with their end grain.

A root maul is not something you can buy. Find a tough hardwood tree about five to seven inches thick at the ground. Dig away the dirt, chopping the roots as you go. Once it's uprooted, clean it well and chop it into shape. Shape the wood while it is green and soft, but let it dry before you put it to hard use. For a two-handed maul, the head might be about ten inches long and the handle twice as long as that.

As long as both pieces bend equally . . .

. . . the plank can split straight.

Froe

You'll need a smaller version of the root maul to use with your froe. The L-shaped froe leads two lives as a splitting tool. As you drive it into the wood,

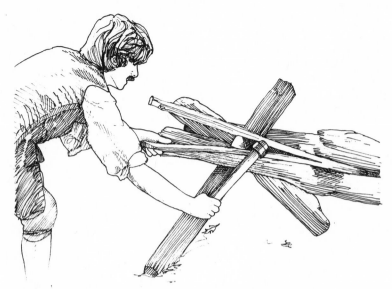

RIGHT: *Bend the thicker side against your knee to keep the split straight . . .*

LEFT: *. . . or use a riving break made from the crotch of a tree.*

Push down on the thicker side to bring the split back to the center.

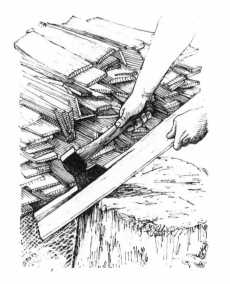

Split off the sapwood with the broad hatchet.

it acts as a wedge. Once sunk in the end, when you pull back on the handle, the blade acts as a lever to pry the two pieces apart.

You'll need to set the wood into a notch of sorts before you pull back on the froe, because the splitting action works only if the wood can't twist or skid away. You can also do this with your body, inclining the wood away from you while bending your knee against it. With the bottom of the wood against the earth, and the middle against your knee, the froe can do its work.

Like a canoe down the rapids, the split is going to go with the flow of the grain. But just as you can guide the canoe from side to side, so too can you guide the split. The split will tend to run to the side that bends more. Bent fibers are pulled and weakened; thus, the thinner side, bending more, tends to get thinner and thinner, and the split "runs out." You can direct the split by deliberately bending one side more than the other. With the froe in one hand, your free hand does the steering. Set the thicker side against your knee and pull back on the top. Flip the piece around and bend the other half as necessary, keeping the split running down the middle.

On long splits, you can stand on the piece and pull up on the thicker end. When you need to do a lot of such work, though, make a riving break from a closely forked limb. The gap should be less than a foot or so—and the more parallel the branches of the fork, the better. The fork needs to lie horizontally, somewhat more than knee high. The bottom of the fork can just sit on a stump, but the far end gets two timbers set through it in an X. The X wants to collapse, but the fork constrains it—like this: ·X·.

Now you can push down on the froe handle, with the sides of the riving break replacing your knee and the ground. You can more easily steer the split, because you can push down with your weight on the thicker side. Flip the board as needed to keep the thickening side down and the split running true.

Repeating this thousands of times in a day's work, the countryman shingle maker exploits the planes of weakness in the wood, and a tree becomes a roof. The froe is the essential wedge of the shingle maker. Shingles fresh from the froe may need trimming, a job for the hatchet's edge.

LEFT: *A straight-bitted broad hatchet does well on narrow pieces . . .*

RIGHT: *. . . or on convex surfaces . . .*

. . . but hollowing calls for a rounded bit.

First, though, why do the shingles need trimming? Oak shingles, split radially from the log, carry a band of white sapwood on one edge. Sapwood is the living wood of the tree just under the bark. After a few years, as more fresh wood has grown outward, the aging sapwood builds up natural preservatives and becomes heartwood. In some trees, like hickory and sycamore, the heartwood has little color change and gains hardly any resistance to decay. In other species like oak, cedar, and walnut, the color and durability change is profound. Sapwood, however, has no resistance to rot. In oak shingles exposed to the weather, it has to come off with the hatchet.

Hand Axe

Wedgelike, the hatchet can split off the sapwood with one or two strikes. It can also finish that surface with its edge. This is hewing, working close to the margins with waste pieces too weak to open a split ahead of the blade. If you want to cut deeper than a quarter inch or so with the hatchet's edge, just score the waste wood with regularly spaced angled cuts. This prevents the waste wood from building up leverage. We're moving away from splitting and toward cutting. The grain is taking less control—we're taking more.

Hatchets have a divide of their own. Side or broad hatchets are beveled only on one side. The flat side is in line with the direction of the blow, and the bending and splitting force is directed to the waste side of the work. For more precise control, you choke up on the handle and steady the blade with an extended finger or two. If you need to shape a concave surface, you work with the bevel down and the flat face to the waste side.

Broad hatchets are usually handled with the bevel to the right, but George Sturt, in his memoir of work in an 1880s wheelwright's shop, recalled the left-handed side axe belonging to one of the wheelwrights in his father's shop: "Other workmen might be annoyed by apprentices or ignorant boys using their sharp axes; but you didn't do that twice with George Cook's axe—it was too dangerous a trick. Why did the confounded tool, albeit so keen-edged, seem

Chop holes through fence posts with a narrow-bitted mortising axe.

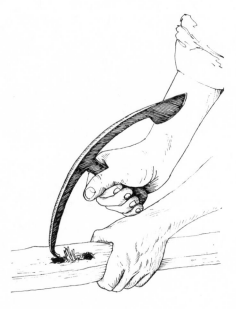

The hurdle maker's twibil.

to avoid the hard wood and aim viciously towards your thigh, or try to chop your fingers off?"

The broad hatchet, with its single bevel and little lengthwise arc to the edge, excels in places where the wood is narrower than the blade is wide. But strike it into the flat surface of a broad timber, and you'll see why you need a common hatchet—beveled on two sides with a curved bit that will penetrate the wood, just like a felling axe. With it, you can chop two cross-grain notches into the face of a split-out poplar plank and then strike with the grain to split out the waste between them. Quickly you can rough in a wooden bowl, ready for refinement with less risky tools.

Mortising and Post Axe

Timber too small for shingles may be fine for fence posts and rails. If you narrow and lengthen the bit on an axe, making it more like a chisel, you can work deeper into the cross-grain of a timber, excavating a deeper and narrower slot than is possible with the hatchet. The mortising axe is sometimes called a post hole axe because, with its shorter handle, you can notch and split out the holes for the rails to pass through fence posts. Single-bitted mortising axes look like felling axes with a Pinocchio nose.

Do you swing a mortising axe or strike it with a mallet? If a tool is made with enough mass to swing it, that's usually what you do with it. Mass is not an asset for a tool that you strike. A struck tool needs to be heavy enough to hold up, but not so heavy as to waste energy overcoming its own inertia.

Double-bitted, T-shaped mortising axes called twibils are always swung. One of the two bits will usually be larger or at right angles to the other. Twibils may be straight in the top of the T or curved to match the arc of the swing. They are powerful tools for roughing in—but challenging for exacting work. It's best to sit on the timber or raise it on trestles to waist high—and always be mindful of a forehead-twibil collision on the backswing.

A smaller tool also shares the twibil name. Five-bar gate hurdles are not so familiar to Americans, but in the British Isles these portable fence sections controlled the grazing of sheep within larger spaces bounded permanently by dry stone walls and hedgerows. The hurdle maker's twibil helps cut the mortises through the posts. It's an odd way of working; you slice the wood with the knife end, and then tear out the wood between bored holes with the leverage of the hooked end—a lever-edge kind of a tool.

Adze

Tusser mentions the countryman using an adze to "make trough for thy hogs." The adze can hollow a hog trough, but the adze is also a faster-working alternative to the plane. In 1678, Joseph Moxon described the adze being used "to take thin Chips off Timber or Boards, and to take off those Irregularities that the Ax by reason of its Form cannot well come at; and that a Plane (though rank set) will not make riddance enough with."

The common adze is a single-bevel tool worked with the flat face against the wood. This flat face curves gently back from the edge, matching the swing of the tool. While one hand powers the tool, your other hand anchors the end of the handle against your body. You adjust the depth of cut with your body position, while, down on the business end, the flat face hammers ever so slightly against the previously worked surface and guides the edge into the unworked wood. In experienced hands, it is dead accurate and leaves behind a smooth, lightly scalloped surface.

The edge slices off thin chips, more like broad shavings. You can work with the grain, across it, or diagonally. There is always a chance that unwanted wedging action would cause the wood to splinter ahead of the edge. For extra smooth work, therefore, some workers set their foot on the surface and swing the blade to cut beneath it. The wedge action is stymied because the worker's shoe holds down the wood until the edge slices it off. As one writer noted, "It is fearful to contemplate an error of judgment or an unsteady blow."

To keep its razor edge, an adze is usually hardened to the same degree as a chisel—too hard for a file. Since the adze has to go on the grindstone now and then, the handle has to be removable. If the handle were wedged in place like that of an axe, it would always be in the way as you tried to grind the bevel.

Carpenter's adzes are rather light tools, worked at high speed removing thin chips that bend up and allow the edge to cut smoothly without splitting. There are also adzes that are intended to split and chop as well as shear a surface. These railroad plate-layers adzes have a thick, heavy poll (the head opposite the edge). Intended for work on railroad ties, their inertial mass can pop out big chips of wood. This mass also makes them tiring to use with rapid, precise blows—but that's not what they're made for.

The shipwright's lipped adze will do all that a carpenter's adze will do, and excels at cutting a trench across the grain of a timber. Without the lips, trenching would be rough edged and much harder, as the splinters would be constantly entrapping the adze. Shipwrights, unlike carpenters, very often work in an arc in front of them. You'll find shipwrights using adzes with three-quarter length handles at close quarters.

A final technical term for adzing: When a timber has been worked over 50 percent of its surface, it is known as half-adzed. (Of the 23 known woodworking puns, a fair share involve the adze.)

Bowl Adze

Troughs suit hogs just fine, but we prefer to call them bowls. A bowl adze works inside a sphere, a world concave in all directions. The adze head must fit within this curve, as must the swing of the handle. A bowl adze shaped into a curve both along its length and breadth can reach in and chop hollows—as when making wooden bowls or scoop shovels. Such an adze will have the bevel on the inside, like a common adze. A bowl adze head can also be straight along its length, but then must be beveled on the outside (convex) edge to give the tool clearance along the arc of the swing.

Trimming the giant dovetails of a cider press with a carpenter's adze.

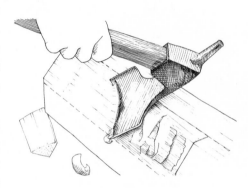

A shipwright's adze gives you clean cuts across the grain.

The bowl adze.

Gutter adzes are similarly shaped, two-handed versions of the gouge-shaped adzes used to hollow long wooden rain gutters. They cut the inside of a cylinder, working straight along the length of the cut. They can have long handles, be beveled on the inside, and be curved only slightly along their length.

A bowl adze is usually an intermediate tool. Except in shallow work like chair seat hollowing, the hatchet roughs in the hollow, the adze comes next, and planelike shavers finish the task. Bowl adzes are usually locally made and often adapted from straight-bitted adzes. Should you propose to change the bevel side on an adze by heavy grinding, check to see that the blade is not laminated, with the steel only on the outside edge. Grind through the steel and you will be left with an edge of wrought iron that will bend over after two strikes.

Firmer Chisel and Gouge

Each blow of the axe or adze combines both guidance and power in the same stroke. With the glut and maul, guidance becomes a separate step from the power stroke. You place the wedge where you want it, then drive it in. You still have to hit the tool with the mallet, but the edge is already on target. You have taken yet another step away from the craftsmanship of risk, toward the craftsmanship of certainty.

A firmer is a broad chisel that you can drive with a mallet, as opposed to a paring chisel that should only be pushed. The name comes from the French, *fermoir*. It's been a confusing name, but Salivet's 1792 definition of the tool clears things up: "The fermoir, or clasp, is a chisel whose steel is gripped between two outer layers of iron. This gives it great strength, and requires that it be sharpened with two bevels. Carpenters employ it to outline rough works. They run from one to three inches broad."

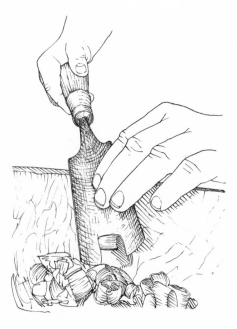

The old fermoir chisel had a double-beveled edge.

Clean up after the bowl adze with a gouge.

Carpenters hardly ever see double-beveled chisels now, but when steel was more expensive and less reliable, it was a good idea to have it sandwiched between layers of iron. These days, what we call firmer chisels are single beveled. If they are laminated, the steel is on the flat face, just as with paring and mortising chisels. Sculptors and carvers still use the double-bevel chisel, but it is all steel. In any case, after chopping something like a bowl to rough shape, you can finish the convex outside surface with a firmer chisel and mallet.

A mallet runs about 15 inches long, including the head, but you can choke up on the handle for lighter work. In theory, the faces of the mallet angle in a bit to converge at the user's elbow. The intent is to keep the mallet face square to the chisel head at the end of the stroke, but mallets generally last long after the bevels have worn away. Live oak, hickory, and beech are common mallet woods, but any dense, split-resistant stuff will do.

Smoothing the inside of the bowl calls for a gouge. The gouge equivalent of the old *fermoir* would have the bevel on the outside of its curved edge. With this bevel riding against the wood, this out-cannel gouge can cut the inside of a curve, diving in and out of the surface.

When the gouge has its bevel on the inside of the curve (the in-cannel configuration), it tends to cut straight or dig in. In-cannel gouges work best for straight paring cuts close to a surface on side or end grain. Working with moldings, you might rough in a long hollow with an out-cannel gouge before shaping it with a plane. Then you'd use an in-cannel gouge to trim the ends to fit over another molding at an inside corner. This is the joiner's work, however, and not that of the countryman.

Bucksaw

This is the firewood saw, the fence post saw, the cut-off-a-piece-to-fit-in-the-lathe saw. The frame of the bucksaw allows for a thinner blade, which makes it both cheaper and easier cutting. Intended for crosscutting green wood ranging up to about 9 inches in diameter, the blade, or web, may have simple peg teeth or some version of the modern-designed raker tooth arrangement discussed earlier.

The blade-stretching wooden frame of a bucksaw is under considerable strain when you tighten the metal turnbuckle or the toggle stick in the twisted cord. The pull on the ends of the blade doesn't stretch the metal—it just keeps it from bending. Once the blade is straightened out, the only give comes from the wooden uprights, bending like archer's bows. If you leave the saw under constant tension, the uprights will lose their elasticity as well as trying to bend to the sides in a cupping manner. Your work is done only when you loosen the saw and let the frame relax.

Green softwood tends to fuzz up in the kerf and seize on a blade lacking adequate set. If you set the saw teeth wider to keep it cutting, it works fine—until you need to cut a length of dried-out dogwood and the blade jumps around and tears into your hand. You may want to keep two bucksaws with different sets to use on different woods.

Knife and Crooked Knife

In truth, a knife is all you really need—but once you've acquired many other tools, you've probably forgotten where you left your knife. Knives are, well, knife-edged tools—sharpened on both sides of the edge so that rocking on the bevel controls the depth of cut.

You don't have much control with both hands working out in space, one holding the work and the other holding the knife. When shaping a long peg, try resting your knife hand on your knee and pulling the peg back, sliding it between your knee and the steady blade. On a short piece, like a dogwood hinge pintle, extend the thumb of your hand that holds the wood and lever against it in a rocking motion with the hand that holds the knife.

Although they are solidly tools and not weapons, the crooked knife is at the center of a long-running academic bar fight: Who invented it? Which end is crooked, the handle or the blade? And what about beaver teeth? Head for the exit as fast as your snowshoes will take you.

The versatility of the crooked knife is legend. Thoreau watched North Woods natives making all the parts for their canoes with these tools. Native Americans

Ease the tension on your bucksaw frame when you're done for the day.

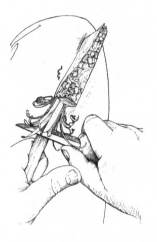

One hand holds your knife as it pivots on the fulcrum of your other thumb.

The crooked knife.

The mighty stock knife.

A shaving horse with a solid head.

The bodger's horse.

are the masters of the crooked knife, but every culture able to feed itself with wooden spoons probably came up with a curved blade to reach into the hollow bowl. Welsh spoon carvers have long used a curved knife with a long handle extended under the arm for leverage.

The familiar North Woods crooked knife, the *mocotagan*, has a turn in the end of the handle for a thumb rest to give you more leverage on the pull stroke. The handles are sometimes wonderfully carved or otherwise decorated. One that I saw had a tiny tintype of a Victorian lovely inset on one side, and a tiny mirror inset on the other — it gets lonely up there in the cold North Woods.

Stock Knife

Since this leverage idea works so well, let's extend the idea by making a knife with one end anchored by a loose hook set into a block and the other end extending to form a lever. With a low-hanging T-handle grip for control, the stock knife was the preferred tool of wooden clog makers. The hooked end limits the length of work that will fit under the blade, but, by definition, clog makers don't work wood much more than a foot long.

A stock knife is supposed to be safer than a hatchet for trimming the edges off short stock — which is probably why I almost lost my thumb to one. The shape of this particular stock knife abetted the mishap. The blade was riding below the line of the hook and handle. Properly configured, with the blade riding at or above a line drawn between the hook and the handle, it is dead stable.

Shaving Horse

Leverage is a handy thing. In the froe, it splits the wood. In the shaving horse, it grabs the wood. The shaving horse is a foot-operated vise that allows you to hold a piece of wood while you sit and shave it with a drawknife or other tool. The earliest image I know of a shaving horse is in a 1500s-vintage German mining textbook. The horse is fully formed, with accompanying drawknife — more evidence of its German origins. Even in the New World, the shaving horse is commonly called a "schnitzelbank," from the German "cutting bench."

The beauty of the shaving horse is that the more you pull with your arms, the harder you must push with your feet. The harder you push with your feet, the harder the jaw grabs. The grip of the vise may be instantly released by removing foot pressure, allowing the piece to be quickly repositioned.

Bodger's Horse

Although grounded in the same principle, there are two sorts of shaving horses. The more common one is the solid-headed horse used on the Continent and in America, as well as by coopers everywhere. But in British woodland crafts, they more often use a bodger's horse. Here, the foot lever and grip is a rectangular frame fitted around the bench and the sloping work surface. The bodger's horse is shorter and lighter—a good thing, because it was carried into the beech forest by the bodger. There, the bodger split billets of beech, shaved them down, and turned them into chair legs on a spring-pole lathe. At the end of a season, the bodger carted his legs to town and sold them to chair makers. Never made a chair, just the legs.

Drawknife

Combined with the shaving horse, the drawknife lets you put all the strength of your arms, legs, and back into long strokes with a razor-sharp blade. The handles that you use to pull the knife along the surface also give you precise control over the angle of the blade, allowing you to cut deeply or finely. The drawknife is still a free and open blade, often taking a middle role in creating a shape—the piece first being chopped, then drawknifed, then planed to final smoothness. The work goes quickly. Sitting on a shaving horse with a good drawknife in your hands, you can get to talking with someone, look down, and find that you've made a chair by mistake.

Unlike pocketknives with two equal bevels, and unlike chisels with one flat and one beveled edge, drawknives often split the difference. The upper side of the blade may have the only visible bevel, but the apparently flat side often carries a slight bevel—just enough to give you some depth control. New drawknives are often dead flat on one side, and when the flat is against the wood, you have control only when cutting convex surfaces. Flip the drawknife over to cut with the bevel down, and you again have control over the depth of cut. Drawknives are so different in their bevels, curvature, and handle angles that you need to get to know them individually.

If you hold it diagonally to the direction of the pull, any drawknife can work with a

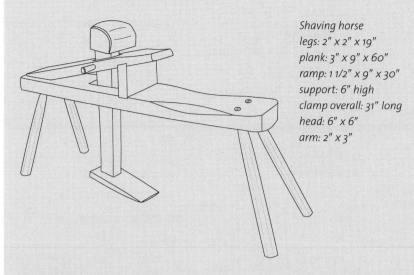

Shaving horse
legs: 2" x 2" x 19"
plank: 3" x 9" x 60"
ramp: 1 1/2" x 9" x 30"
support: 6" high
clamp overall: 31" long
head: 6" x 6"
arm: 2" x 3"

Two Shaving Horses

This horse lets you shave
    with the drawknife's keen edge,
as you sit on its back
    split from logs with a wedge,
which you rive out by breaking
    the grain that is weak
that flows through the oak
    like gray hair in a streak.
The legs fit in holes
    in the plank like a stool,
each fixed with a wood wedge
    that follows this rule:

When setting the legs,
    fit the wedge 'cross the grain
of the plank that you sit on,
    its strength to maintain.
When shaving, you'll see that
    the harder you draw,
the more your horse bites
    on the wood in its jaw.
And though your fine horse
    won't win any races,
your seat on its back
    is the best of all places.

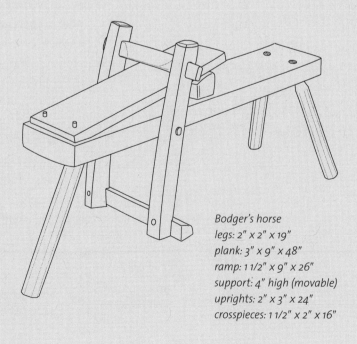

Bodger's horse
legs: 2" x 2" x 19"
plank: 3" x 9" x 48"
ramp: 1 1/2" x 9" x 26"
support: 4" high (movable)
uprights: 2" x 3" x 24"
crosspieces: 1 1/2" x 2" x 16"

Hold the drawknife slightly askew.

The curved shave works inside a hollow.

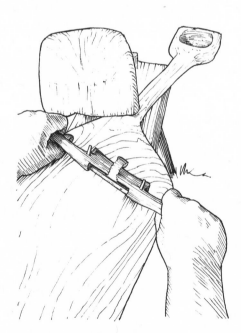

Rock the wooden-bodied spokeshave slightly forward as you work.

shearing cut. This lowers the slope of the upper bevel—a mixed blessing. A lower bevel slope makes a sharper edge, but also a thinner wedge. The blade may cut better, but the shaving may not be bent enough by the bevel to break it before it builds up enough leverage to split the wood open ahead of the edge. Different combinations of wood moisture and grain structure make this a matter of each moment's judgment—part of the reward of working with these wonderful tools.

Coopers use convex knives to hollow the insides of their staves. Convex drawknives with high handles called inshaves can work the hollows of a Windsor chair. I call single-handled versions scorps, but this is a good time to recognize that the names for many of the tools change over time and distance. My moot may be your nug, our neighbor's rung engine, and someone else's thole reamer. Very confusing, considering that we're all talking about the same witchet.

Spokeshaves, Wooden and Iron

We've split and chopped and adzed, working with tools that could also be used to storm the castle. You wouldn't do that with a spokeshave. This is the first move away from the wildness of the open blade, now the depth of cut is fixed in the tool. The spokeshave can smooth, but it can't do sharp transitions or tight inside curves. We're working with a plane now, and risk is getting left behind.

The name suggests that the spokeshave was primarily a wheelwright's tool, used for smoothing the rounded surfaces of wooden wheel spokes. But the word "spoke," in an older sense, can mean any long split-out billet of wood. So coopers, shoe-last makers, and anyone working with wood from the wedge might lay claim to the spokeshave's edge.

Wooden-bodied spokeshaves use cutting irons with the flat face against the wood. The ends of the blade turn up into tangs that make a tight fit into matching holes in the body of the tool. You adjust the depth of cut by tapping either the protruding ends of the tangs or the extremities of the blade on any handy hard surface. Later models use thumbwheels to adjust the depth of cut, but the blade arrangement is the same.

In use, wooden spokeshaves must bear entirely on the narrow ledge of wood in front of the cutting edge. It takes a little time to get used to working "rocked forward" rather than with the flat of the blade against the wood.

Iron spokeshaves use a different configuration of stock and blade. An old shop textbook advises, "The spokeshave is practically a short plane with handles at the sides, and in using it the aim should be, as with the plane, to secure silky shavings of as great a length as the nature of the work will allow."

Planes have been around for 2,000 years, but only when malleable cast iron became available in 1860 was it practical to make this "short plane." If you can make the body of a plane in cast iron, it can be much shorter than is possible if it's made of wood. The basic plane or iron spokeshave consists of an iron (the blade) sharpened on one face to about 30 degrees, mounted, bevel down, at 45 degrees in the body. Like any plane, these tools allow you to work with their whole bed riding on the wood. They come in a thousand patented configurations and contours, and can handily combine a concave and straight cutter in a single tool.

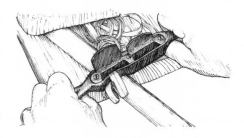

The iron spokeshave rides more like a plane.

Iron spokeshaves have some advantages over wooden-bodied spokeshaves. They are sturdy and longer wearing, but don't let the pleasures of the wooden spokeshave pass you by. Wooden shaves adjust with a one-handed tap on the bench, and there's nothing like the feel of the warm beech stock in your fingers as gossamer ribbons of end grain flow from the low-angle blade.

The Devil, the Travisher, and the Forkstaff Plane

The devil and the travisher share the form of the wooden spokeshave. The devil has been around forever—ever since we broke open that first stone and cut ourselves on the sharp edge. The chair maker's devil is a scraper in spoke-shave form. As with any good scraper, it doesn't scrape as much as it shaves. The vertical iron is ground at 60 degrees and may be lightly turned to hook forward with a few strokes of a hard steel burnisher. In spite of its name, the chair maker's devil is a sweet-cutting tool and can shave any grain perfectly smooth.

The travisher is now the name for a high-handled spokeshave with a convex iron and bed used for smoothing Windsor chair seats. My old sources also tell me that a travisher is an ordinary-looking, straight spokeshave for spindles. The curved spokeshave now called a travisher was known to English chair makers as a smoker-back-hollow-knife.

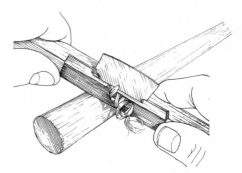

The devil—a scraper mounted in a wooden stock.

Here's my theory. Smoker-back armchairs never became a part of the Windsor chair making revival. The curved shaves remained, however, still burdened with the clunky and now meaningless name smoker-back-hollow-knife. What to call them? Well why waste that leather-jacketed cool name, "travisher," on a regular-looking spokeshave? Sorry pal, the cool name now goes to the guy with the round bed.

It could also be that my old sources are askew. It happens. Still, a few years ago, some modern timber-framers, working with huge beams, began calling themselves joiners instead of carpenters. If this catches on, a thousand years of meaning will have shifted in our lifetime.

A short plane can follow curves and contours; often though, you want a tool with even less freedom. When you want to shape a straight piece, a longer bed will carry the iron over hollows, shaving only the hilltops until the high places are made plain.

Among countrymen, there's usually someone who supplements his farm income by making rakes, hayforks, and grain cradles for his neighbors. The forkstaff plane may be the only plane in his workshed. Concave in cross section but long in the bed, it will plane the staff of a hayfork round and straight. The forkstaff plane is the country cousin of the molding planes of the joiner, its iron contoured to match the shape of the finished product.

Use a forkstaff plane for long, straight handles.

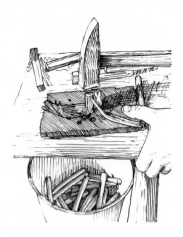

Drive the peg almost through the dowel plate, then start another on top of it to knock it out.

Pull long, small rods of split wood through the dowel plate . . .

. . . shaving it through progressively smaller holes.

Peg and Dowel Plate

A wooden rake needs teeth, wooden pegs driven into holes in the rake head and wedged. When you want to drive a peg into a hole, a certain degree of misfit helps the grip. Roughly rounded, octagonal, or even square pegs grip better, but too much of a square peg risks splitting the round hole.

When round is what you want, you can drive the square-split wooden peg through a round hole in a steel plate. The steel plate will shear all character from the rough peg, allowing only cylindrical conformity to emerge. When making tines for a wooden rake, you set the plate over a hole in the bench, drive the billet almost flush with the plate, and then start another billet atop it to drive the first one all the way through—shooting down into the bucket set below the bench with a satisfying plunk.

The rough wooden peg will shear more easily if it is still fresh and soft from water in the cells and cell walls. But as this water evaporates, the wood will shrink and you'll have a peg that drops easily through a hole that it once had to be driven through. The hole drilled in the steel plate is going to stay the same size from day to day—but not so the peg you drive through it. That's why a dowel plate you drill for yourself may do better for you than a store-bought one with fewer gradations in size. You can drive green hickory pegs through an oversized hole, let the pegs dry and shrink, and then drive them through the final-sized hole some weeks later.

Mild steel works fine for a dowel plate. You can always drill out a dulled hole to the next larger size. Counter-drill on the underside of each hole with a bit that is the next size larger, leaving about 1/16 inch of the original diameter remaining. If you want to ensure that the pegs are straight as well as round, mount the plate over a thick wooden block with slightly oversized holes bored beneath each hole in the plate. The long holes through the block will hold the pegs straight as they emerge.

There are not many items outside of hay rakes and scythe cradles that call for them, but sometimes you need long slender rods of straight-grained hickory or the like. You can't drive these through the dowel plate—you have to pull them through. As always, the wood has to be split, not sawed, so that the grain stays intact.

Pulling is harder than hammering (imagine pulling a nail into a board instead of driving it) so work these rods down through a series of gradually diminishing holes while they are still fresh. Then let them shrink to their final size.

Tenon Cutter and Spoke Pointer

If pegs had shoulders, they'd be tenons. Tenon cutters were once an obscure wheelwright's tool, but the popularity of rustic furniture has brought a new generation of designs. For the wheelwright, the tenon cutter, also called a hollow auger, quickly shaped a precise, round tenon on the end of a spoke, ready to fit into a hole bored through the felly, a section of the circumference of the wheel. For the rustic furniture maker, hollow augers quickly terminate nature's shapes with simple symmetry. A curving branch of ironwood becomes the arm of a rustic chair with a perfectly round tenon to fit into the perfectly round mortise bored in the back post.

Tenon cutters work on end grain and may be engineered with bevel-up or bevel-down cutters. The wheelwright's tool is designed to leave a square shoulder to take a wagonload of pressure. The furniture maker's tools are shaped to leave beveled or rounded shoulders for a more gradual transition from nature to culture.

A tenon cutter won't start on a bare, square end. It requires just enough wood already fitting in the central hollow to keep the orbiting cutters from wandering off. Like a big, blunt pencil sharpener, the conical spoke pointer fits over the sharp-shouldered spoke and planes away an ever-widening shaving. You stop when the end is just a bit smaller than the tenon you want to cut. Smaller versions forged from a single piece of steel can point a dowel, or just take off its shoulders.

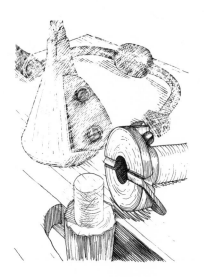

The tenon cutter or hollow auger.

Rounder Plane and Taper Auger

If the spoke pointer is like a pencil sharpener, the rounder plane is like an endless pencil sharpener. The stick gets shaved down to a certain size and then passes through a hole in the tool. You can work as far as you want, making just a tapered tenon on the end of the stick, or unwrapping shavings down the whole length.

A rounder plane often gets named after the item it shapes. Ladder makers call it a rung engine, but rake makers call it a stail engine, after the name for a rake handle. Boat builders call it a spar shave. When made in two parts with hand screw adjustments, the rounder can shape different sizes or, with continual adjustment, shave a taper. In America, these adjustable rounders are called witchets.

It's slow going, though, so try to do most of the work with a drawknife. The wood will emerge very rough if the rounder plane iron is set too deep, and the rounder plane will be very hard to turn if the iron is too shallow. Candle wax rubbed in the opening is a great help. A rounder plane will follow a curved stick very well, sometimes too well. If you leave a longer, noncutting, cylindrical passage on the trailing end of the hole, the rounder will tend to straighten a crooked path.

A cooper's bung borer shapes the tapered hole to make a rounder plane.

Taper augers cut conical holes for coopers, ladder makers, wheelwrights, and Windsor chair makers. They are side-cutters, and should only be sharpened on the inside edge. Starting with a taper auger, you can make the taper rounder to shape the matching tapered tenons. The positive makes the negative.

Scraper-based taper reamers are simple to make. Turn a tapered hard wood body, saw it down the middle, and insert a thin, tapered steel blade. This square-edged scraper working within a circle will still have clearance, but you can make it cut faster by filing a relief angle on the trailing side. Adjust the exposure of the blade by adding spacers between the end of the saw cut in the wooden body and the scraper blade. Marvelously, this type of scraper blade is self-centering.

Use constant pressure to keep the nose, or shell, auger cutting.

The screw-pointed, spiral auger draws itself into the wood.

A brace and bit makes the boring continuous.

As violin and Windsor chair makers know, a long, tapered tenon gives tuning pegs and turned legs a tenacious grip.

Augers

Boring tools move the action from outside to inside, and give all these pegs someplace to go. The T-handled auger is "complete in itself," acting like a rotary chisel to shave out the wood as the tool progresses. Intended for large holes where you really need the leverage, the action is necessarily intermittent as you reposition your hands for the next turn.

The marks left in old buildings by auger bits can help you determine their age. Just be sure that the hole is from the original construction. Many a timber is perforated by generations of auger holes — perhaps to peg up a shelf, but apparently just for fun as well. The screw-pointed, spiral-shanked auger that lifts the shavings out of the deepening hole was available in the late 1700s but did not become common until about the 1820s. Before then, shell or nose augers required a centering, starting hole gouged into the surface of the wood before they could begin cutting. Once down into the wood, these augers require steady pressure but cut well enough, leaving the shaving intact as a single, springy spiral.

Both shell and spiral augers have chisel-like cutters that become worn with use. If these cutters are rounded over on the upper face, they won't cut well. If rounded over on the lower face, they won't cut at all. File the lower faces of the cutters just enough to bring them flat to the edge and do the rest of the sharpening on the upper faces. The auger bit files with "safe" edges are too small for big augers, so mind the corners of your file when working near the lead screw.

Big T-handle augers often have vertical cutting lips that sever the ends and free the chip, much like the upturned lips of the shipwright's adze. Sharpen these only on the inside to ensure that they cut a hole bigger than the body of the auger. Encounters with badly worn augers remind you to clean the dust and dirt off your timbers before you put your tools to them. Tools fare badly cutting through embedded grit.

Brace

The crank brace that allows you to bore a hole with a continuous motion seems so obvious now, but Europeans thought of it only about 600 years ago. Like the T-handle auger, the brace is essentially a rotary lever — the offset of the crank makes your hand travel farther in exchange for added strength. A brace with a compact throw radius of less than four inches will quickly bore a series of half-inch holes to start a mortise, but will test your strength when cranking a one-inch bit through hard maple.

Used metal braces with ratcheted, self-centering jaws and ball-bearing heads are cheap enough that you can have several. Some wooden braces, however, go for many thousands of dollars. These ebony and ivory braces were made as presentation pieces, but even a common beechwood brace still carries a certain mystique. We know in our bones enough about wood grain and strength to find the offset shape of a wooden brace unsettling in itself.

Wooden braces commonly use a spring chuck—a tapered square socket with a button-released catch. Bits had the matching square taper and a notch to receive the catch, but the shape of the tapering square varied enough that you were well advised to buy them in matched sets from the same maker. A century or more of jumbling in time's junk drawer now resigns us to wobbly fits between braces and bits.

Spofford-chucked braces split the difference between panache and utility. Developed in the 1860s, about the same time as the familiar Barber double-jaw chuck, a Spofford is an iron brace with a thumbscrew-tightened split socket. Spoffords center well on any taper square bit and retain the clean lines of a wooden brace, lacking only the seldom-needed up-against-the-wall advantage of a ratchet jaw.

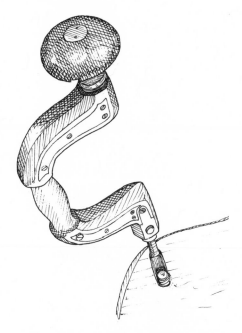

Chair makers prefer the rounded spoon bit, here in a wooden brace.

Bits

If we take an outside-beveled gouge and begin twisting it into a board, it will dig deeper and deeper, making a hole. Shell and spoon bits work just this way. Soup would run out the end of a shell bit, but spoon bits are turned up on the end just like their namesake. They cut on their rims and have to be scraped sharp on the inside with the ground-off corners of a triangular file. Chair makers greatly prefer them. They can begin a hole at right angles to a surface, then allow you to tilt the bit like a ball in a socket to continue boring at a new angle. They don't have a lead screw to split the wood, and can bore deeper before poking through the other side. An old spoon bit will often have marks filed across its back, depth guides for the chair maker who owned it before you. Tapered square tangs make these bits interchangeable, but many coopers and chair makers used spoon bits, each permanently set into an individual wooden brace.

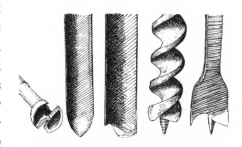

A Forstner bit, a pointed spoon bit, a nose bit, a Jennings spiral bit, and a center bit.

A nose bit keeps the semi-cylindrical form but adds one upturned cutting lip. Unlike the larger, T-handled versions, these smaller bits will usually start themselves without you first having to gouge a centering pocket. You can sharpen the lip of a nose bit with a small file. Even more filing can reshape broken or very worn nose bits into shell bits, extending their boring lives.

Center bits have about as elegant a design as you will find. The central pike drills in and holds the outer spur in perfect circular orbit around it. The spur knifes down into the grain and defines the perimeter of the hole. The blade enters and, like a rotary chisel, sweeps out shavings of wood as the bit is forced deeper into the hole.

Sharpen only on the inside of the center bit's spur or nicker.

Center bits look too simple to perform as well as they do. They don't eject shavings in deep holes, but they cut perfectly clean and round. As with any bit, you can make a cleaner exit hole by clamping waste wood to the far face. More often, when the pike begins to poke through on the other side, you flip the piece over and bore back to meet the original hole. Take it easy, though. The

The Jennings-pattern spiral bit has two nickers and a flat twisted body.

With no spurs in the way, a Gedge pattern bit can easily start at an angle to the surface.

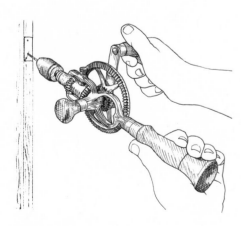

Cast iron parts make the hand drill affordable, but fragile.

center bit can poke through before cutting that last part, leaving a web but not enough wood to keep the bit centered.

Spiral bits evolved from the center bit in the late 1700s but did not become common for decades. Walter Rose, in his memoir of an 1880s English carpenter's shop, recalled that his father's tool chest had specially prepared places for center and shell bits, but none for the newly introduced spiral bits.

Auger bits are usually stamped with a number indicating their diameter in sixteenths of an inch. Bits are also made with lead screws giving different rates of progress for fast or fine work. The lead screw on any give spiral bit draws the cutters into the wood at a constant rate, so you can keep track of the depth of a hole by counting the turns as you progress. The lead screw also acts as a wedge, so it's wise to clamp the sides of narrow pieces.

Center bits and spiral bits benefit from the safe, untoothed edges of the auger bit file. The downward cutting nib or nibs should be filed only on the inside. The cutting lip only needs flattening on its gently angled bottom face—unless the nibs are getting too short. Over time, the nibs wear down, and if you don't file back the lower faces of the cutting lips, the nibs can become too short to shear.

Every bit drawer has its dogs and cats. The exotic Gedge pattern auger curls up its cutting lips like a waxed mustache. This breed digs in quickly, biting into end grain. Forstners have a century-long pedigree. Unlike other bits, Forstners track on their perimeter instead of their centers. They can dig flat-bottom holes partially overhanging a shoulder or deeply overlapping other holes. They're generally shallow-cutters and not fitted with a square shank for a brace. These factors, plus their greater cost, make them occasional rather than everyday tools—just like the homely expansive bit that can almost cut any size hole. As they say down at the Possum Lodge, "If a tool can't be handsome, it'd better be handy."

Drills and Gimlets

In the mid-nineteenth century, the advent of malleable cast iron put into workmen's hands mechanisms that once existed only in great windmills. Each turn of the large gear on the hand drill gives multiple turns of the smaller gear and its attached drill bit. An increase in speed is always made at the cost of torque, but the difference in the size of the circle described by the hand crank and the diameter of the drill more than compensates for this.

Many drills also have a smaller driving gear concentric within the larger one to use with larger diameter bits. Larger bits also need more pressure and a steadier bearing to keep their broader cutting edges shearing the wood or metal beneath them. The breast plate on heavy-duty hand drills lets you employ the mass of your body, freeing one hand to hold a side handle while the other hand turns the crank.

Reciprocating drills such as bow drills, pump drills, and push drills that work on an Archimedean spiral do well in a narrower range of the density-to-diameter continuum. In other words, they bog down in softer wood and lack the torque for larger holes. But when it comes to making pilot holes for small screws, I'll reach for the push drill or the bradawl. This last is a tool like a sharpened screwdriver that is first pushed in across the grain and then twisted to push the severed fibers away.

On the business end, straight-fluted and spiral bits are much the same, two shearing edges radiating in a conical point. The same pattern that drilled rivet holes in Henry VIII's armor helps you hang a curtain rod today.

Then there are gimlets. Who looks on their gimlets with affection? You may not appreciate the speed and ease of a hand drill until you have had to put a few miles on a gimlet boring holes for a pegged shingle roof. No more than a T-handle on a small drill bit, they are a nuisance to sharpen, easily broken, kill your arm, and split the wood. Other than that, they're great.

Gimlets come in various configurations, most often screw-pointed with either a crescent moon or spiral-fluted cross section. They cut on the sides as well as at the point, and any re-sharpening has to be confined to the interior hollow. A scraper ground from the hard edge of a file will do, as will a round slipstone. I've read that you can sharpen spiral-fluted gimlets by boring a hole with the gimlet, putting oil and emery powder in the hole, and then turning the gimlet back and forth within it. I don't know anyone who bothers.

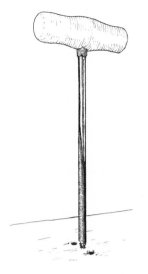

A shell gimlet.

Steam Bending

But everyone likes steam bending—it's captivating to watch. I've seen hickory snow skis bent after a hot water bath, seen walking cane handles bent in tight circles after boiling in a metal drum. I've seen long, thick pine planking for ships pulled from a huge steam box and twisted and bent into place by six men, still steaming as they clamped it to the massive ribs. Even if it's just tiny strings of holly inlay curled around a hot brass rod, it always seems magical—when it works.

Bending wood works sort of like bending your arm. Muscles can only pull. They bend your arm by contracting on the inside of the bend. Keep that inside muscle contracted, and your arm stays bent. It's the same with steam bending wood. Heat softens wood fibers and the stuff that holds them together. Bend the wood while it is hot and the fibers inside the curve will compress. Keep the wood compressed until it hardens up again and it will stay compressed. It will hold the curve.

Or, it may break. When you bend wood, something has to give. All the wood is bending, but the convex face is under tension and the concave face is under compression. Either the inner face gets compressed or the outer face breaks apart. A metal compression strap helps wood stay intact for intense bends. With stops at each end spaced at the unbent length of the wood, they ensure that the bend is all compression, and no tension. Things like chair backs and hayforks with gentler bends can do fine without straps, as they have for centuries.

The initial moisture content of the wood influences the success of the bending. Green wood bends easily but springs back unless you let it dry thoroughly. Kiln-dried wood is already so compressed that it just snaps. If you want to bend

Boiling or steaming softens the wood so it can compress on the inside of the curve.

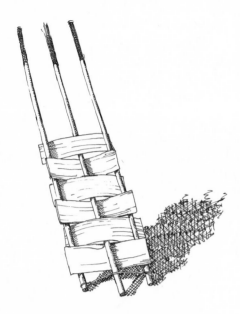

Once the bent wood piece is cooled and dry it will hold its new shape.

Lay out parallel pieces at one stroke.

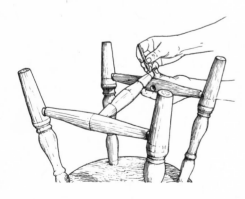

Fit subsequent pieces into the spaces created by previous work.

wood, choose air-dried stock with as much straight grain as possible. I would advise you to work only with riven out wood, but two of the best bending woods are unsplitable elm and sweet gum. The other good benders—ash, beech, birch, hackberry, hickory, maple, oak, and walnut—are best when riven, but can also be sawn from straight-grained, knot-free stock.

Steam gets the heat deep into the wood. The larger the piece, the more time it takes for the heat to soak in. The time required also depends on how hot your steam box gets. I use a teakettle on the woodstove connected by a short radiator hose to a double-walled pine box about four inches square inside and six feet long. The most cooperative wood needs an hour per inch of thickness, while resistant wood gets twice that.

Of course you can also bend in hot water, or just with raw heat. But steam keeps the wood from scorching, so even when dry bending thin wood on a heated pipe, some moisture on the wood helps.

The wood starts cooling and setting up right away, so you have to work fast. The wood will spring back to a certain degree, so your mold needs to be perhaps 10 percent tighter than the bend you want. The wood needs to stay in the jig for several days before you can tie it with string and take it out.

A Few Principles

Whether it's a rocking chair, a post-and-rail fence, or an orchard ladder—items with parallel pieces benefit from gang layout. You simply arrange all the vertical elements side by side, align the ends, and mark the connection points with the horizontal elements across all pieces with the same stroke. Do the same for the horizontals. This is standard practice in joiners' and cabinetmakers' work, but even the roughest split locust fence posts can be marked out for mortising a dozen at a time with the snap of a chalked line, using the same principle of gang layout.

Of course, marking all the pieces at the same time means that you'd better be right about your measurements. So, in addition to confirming all your dimensions before you cut, be sure to stop after cutting that first piece and try it out in place. You can still use the remaining pieces if something is amiss.

At some point you have to make a commitment and cut a piece to length, but the "leave it long" maxim will serve you well. Extra length on a piece gives you something to hold onto, as well as keeping your options open for correction until the last minute. Holding the wood while you work it is no small part of the job. For a common example, imagine that you need to saw a six-inch-wide, two-foot-long board into four equal pieces. If you first cut the board across the grain into a pair of one-foot-long boards, you'll have a tough time holding these short pieces on sawhorses for the lengthwise ripping. But if you reverse the order and cut down the grain first, the long 2-by-3s will be easy to cut to length.

Conversely, when you're making items such as long, narrow moldings with planes, you often want to "leave it wide." The extra width gives you something to clamp in the vise or on the bench top as you shape the molding. When you're done with the shaping, you saw off the narrow molded strip. You may have heard of Native American artists who carve ceremonial masks into the side of a living tree and then split it away only when they are done. This may be based on spiritual rather than mechanical principles, but it sure holds the wood steady.

Leaving a piece long also allows you to fit it into dimensions created by earlier steps. You join three pieces together and then fit the fourth piece in between them. You don't know what the numerical measurement is: you just hold the last piece in place, mark how long it needs to be, and then cut it to fit. Your work grows in an emergent, building process—going with the flow.

White Oak, Black Ash, and Hickory Bark Bottoms

Chairs with bent backs and split splats need woven wood seats. White oak splits for baskets and chair bottoms come from the sapwood of a soft, straight, smooth-barked tree about three to eight inches in diameter. Split the log into pie-shaped sections, and then split off the heartwood as close as you can to the dividing line between the light and dark wood. Shave off the bark and any remaining heart, working it down until you have a piece of clean sapwood that is as wide in the plane of the growth rings as you want the splits to be and as thick as the sapwood is on the tree. From now on, make all splits in the same plane as the annual growth rings.

Take a stick and knock the blade of your jackknife into the middle of one end. Once started, work the split down the rest of the way with your hands. Keep the split centered by putting more bend into the thicker side.

Continue splitting each piece in half, always with the growth rings, until they are as thin as you want. Sometimes the wood on one side of a tree will not split out well, while that on the other side splits just fine. There can be as much variation within a single tree as there is between different trees. If you can't split all the wood right away, leave the log in a shady spot or in a pond until you can get back to it.

When you finish preparing the splits, gather them up and set them aside to dry. If you were to weave them now, they would shrink in width so much upon drying that even the tightest work would loosen up. When the time comes to use them, don't soak them in water or you will get the same kind of shrinking problem. If they seem too dry, just dip them in water for a minute or so to soften the surface.

When the splits are almost as thin as you want them, you can shave them smoother by drawing them between your knee and a knife held vertical to their surface. If the splits are wider than you want, you can cut them to any width with a pair of regular scissors.

Ash splints, like those from white oak, begin with a soft-barked, straight, and knot-free tree, about six to twelve inches in diameter. With ash, instead of directly pulling the wood apart into strips, you first pound on the billet to separate the growth rings so you can then pull them apart easily. Pounding crushes the big cells formed in the spring of every growth ring, while leaving intact the denser cells formed in the summer.

Oak and ash splits make bottoms and baskets.

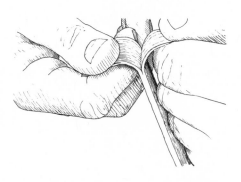

Steer the split down the middle by bending the thicker half more sharply.

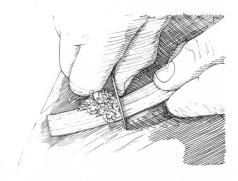

Pull the strips under a knife to scrape them smooth.

Split the green wood down into rough billets and then shave them rectangular with the drawknife, carefully keeping square to the surface just under the bark. Set the billet on a smooth stump and pull it along as you beat on it with a sledgehammer or the rounded poll of an axe. Make this first pounding just hard enough to loosen any short lengths of the outermost and innermost growth rings. Clean these off with the drawknife and then trim the sides to the width of the splits you want. Trim the ends of the billet with your hatchet, or with the drawknife swung like a hatchet.

Now start pounding in earnest. Strike with even, hard blows of the axe head or sledge all the way down the billet's length, compressing the rings like hitting the face of a deck of cards. Turn the billet over and go down the other side. Never strike the billet edgewise to the rings.

The final beating works by shearing the cells crushed by the pounding, just as you would separate playing cards by bending the deck. Feed the billet over the edge of the block and strike at the overhang to rack the layers apart. The growth rings should separate easily by hand, but there may be some hangers needing further pounding or a cut with the pocketknife. Just as in working with white oak, shave the splints smooth by pulling them under your pocketknife.

Shave off the rough outer bark of the hickory, then pull up strips of the inner bark.

Bark strips need to be made in the spring. In the spring, trees rush to be the first out with their leaves, and they need tons of water passing through a fresh layer of big, hollow cells. This fragile layer of rapidly dividing cells makes for easy separation of the bark from the wood. Moonshiners made pipes of hickory bark to run water to their stills — slitting a tree down one side, peeling the bark off in one piece and letting it return to its cylindrical form. Thoreau mentioned a boat on Walden Pond that was anchored by a cable made from strips of hickory bark and, of course, hickory bark strips also make beautiful and lasting bottoms for your chairs.

Look for a hickory tree that runs between six and ten inches in diameter. Smaller trees have thinner bark, sometimes too thin. Try to find a tree with few knots that tends toward the cylindrical rather than tapering heavily. Chop the tree down and cut the top end off at the point where you figure it becomes too small or knotty. I strip the bark off the tree immediately, but I understand that if you let the tree lie for about a week first, the bark will hold its light color instead of turning brown.

Prop the log up on its stump and, with a drawknife, shave off the outer rough, hard, scaly bark on the top side of the tree. Shave just deep enough to get through to the light-colored stuff. Now take your pocketknife and cut with the grain down the length of the tree to divide this underbark into strips as wide as you want.

Pry one end of a bark strip free from the wood and slowly pull it off down the length of the tree. Watch for snags and be ready to cut them free with your knife. Roll the strips around your hand as you go and hitch the end around the coil to keep it from springing loose.

You can weave with the bark just as it is stripped from the tree, but most folks split it into half of its initial thickness and weave only with the inner part. Some people split the bark but use both halves, including the outer, brittle stuff.

The splitting is the same as that used for working white oak. Take one end and work your knife into it lengthwise, giving the knife a twist to open the split. Take half in each hand and work the split down the length of the strip. As always, should one of the sides start to get thicker than the other, pull more sharply on it to bend it more than the other and make the split run evenly again. As the bark coils dry, they turn a deep brown color. Even holding them in your hands you would mistake them for leather.

Seat Weaving

Seat weaving begins with warping the chair, that is, wrapping the initial course of strips in one direction before the crosswise interlacing. Both bark and wood swell when they absorb water and will shrink back when they dry, so dip the splits in water just until they are easy to work.

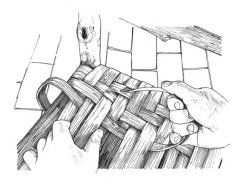

Taper down one end of the first strip and catch it around one of the back stretchers. Bring it around under the front stretcher and loop back, pulling tight as you go around and around. When a strip runs out, make your splice on the underside of the seat. Bark is pliant enough to tie in a knot. Oak and ash will need a hook-and-eye splice or short overlaps lashed with string. With hickory, you may want to keep the outside of the bark facing up and out. When the bark dries, it will crown and make a seat that is more comfortable than, and just as attractive as, the smoother-surfaced but cupped-in inner face.

The herringbone weave goes under two and over two.

When you have wrapped the chair from front to back (or side to side) turn under and around one of the back posts and start the weave. In the herringbone pattern, each weaving strip goes over two and under two warp strips — but begins one step out of phase on each pass. Begin the first pass by going under two, the second pass by going under one, on the third go over two, and on the fourth go over one. Beat your seat tight as you work. Repeat this sequence to complete the weave in long diagonals on a seat to last a century.

Grindstone

Axes, saws, and augers are tempered so that you can sharpen them with a file. On most chisels, gouges, adzes, shaves, and knives, however, a file will skate off the hard edge. These hard-tempered tools must be sharpened with a stone, and even those tools that a file can cut will benefit from the finer edge produced by the whetstone. The whetstone gives the final polish to the edge, but the work begins with the coarse grit of the sandstone wheel.

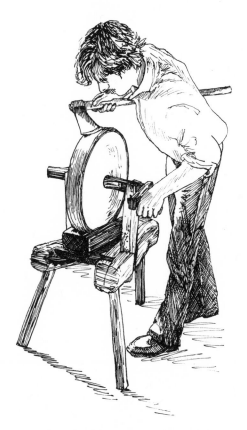

When you acquire a good sandstone grinding wheel, you become the guardian of a fragile treasure. Knock it and it will chip. Let wet wooden wedges expand in the axle hole and the stone will split. Leave it in water and it'll go soft. Leave it in the sun and it gets too hard to cut. Leave it unattended and someone will dry grind a bolt head, leaving a deep groove in the face.

The fragility of these stones has even created controversy. Some people believe that a stone turning away from the tool gives a better edge, but most insist that the advantage runs the other way. Some, like my grandfather, turned

Keep the sandstone wet while you work, but never let it sit in water when you're done.

the stone away from themselves because they just didn't like water being thrown in their laps. The real difference comes down to this, whether the stone is lumpy and bumpy or not.

Turning toward the edge cuts faster, as the wedging action pulls the steel into the stone. Sadly, many sandstone wheels have been left sitting in a water trough at some point in their lives. This irreparably softens the stone at the waterline, leaving two points on the circumference that quickly wear into hollows. An edge tool dipping into one of the hollows tends to dig into an approaching stone, and the only remedy is to turn the other way.

Water is the friend and enemy of sandstone wheels. Dripping from a suspended can or held in a trough below, water washes away the dulled grit and steel particles ensuring a constantly fresh cutting surface. Water cools the heat of friction and protects the temper of the tools. But left sitting partially in water, the stone is ruined.

As with any stone, try to avoid putting grooves in it by constantly moving the tool back and forth to use the whole face. And don't think a sandstone wheel won't cut fast. Especially when turning away from you, a long wire edge can hide the fact that you have ground away a quarter inch of your blade, and that you should have moved on to the whetstone five minutes earlier.

Grinding is the first step of creating an edge. The coarse grit of the sandstone cuts fast but leaves a scratched and sawlike edge. The grindstone does the shaping—the honing comes later. The ideal shape for any given edge tool depends on the steel and the job, but a good starting point is 30 degrees. You can easily eyeball a 30-degree angle by grinding a bevel until it is twice as long as the tool is thick at the end of the bevel slope.

Grinding is reserved for the bevel side, but check the flat face to see if it is rounded over. In planes and chisels, you can't get a proper edge until the flat face is flat all the way to the edge. If the flat is badly rounded over, grind from the bevel side until you cut back to a level surface. In rare cases where a tool is heavily corrosion-pitted on the flat face, you might try to save it by grinding the flat. The flat face is the side with the hard layer of steel that forms the edge—grind it away, and you have a paint can opener.

Whetstone

Honing follows grinding. The more finely gritted stones make finer scratches in the steel, eventually leaving it mirror polished right down to the point where the bevel and the flat face intersect. You don't need a microscope to see this. Look straight on at an edge held in the sunlight—any brightness is dullness. The intersection of two lines is a point, and a point has no width and will not reflect light back at your eye. Intersection is a big word for a tiny place, but that's your edge.

Just as with the sandstone wheel, whetstones need to wear away to keep fresh sharp grit at the surface. Some stones need oil and some need water to carry away the worn particles of grit and steel. Sandstones and Belgian clay

stones are common natural water stones. Arkansas and Washita stones are
natural oilstones.

Olive oil does fine on oilstones, but any oil makes stones pick up all the dust
and dirt in the shop. Lamp oil can help de-gum a sticky stone. Any stone needs
to be wiped clean and kept covered. Natural stones are particularly fragile and
deserve a wooden box to live in.

Start honing with the flat face of the tool lying flat on the flat stone. Work
it in a figure eight pattern over the whole stone until it shows polish down to
the edge. A coarser stone will cut faster if you're not reaching the edge, but
keep the flat side flat. You can't easily flatten on a hollowed stone, so you may
need to level the stone by rubbing it against another one or on a flat surface
covered with abrasive.

Move to the finer stones for the final polish on the flat side and then turn to
the bevel face. Again, start with the coarser stones and work toward your finest.

You have already ground the bevel edge to 30 degrees, and you can continue honing that entire face until it is polished and flat all the way to the edge. Some prefer to reduce the work by raising the angle of the tool a few degrees to hone a second bevel. This steeper honing bevel still needs to be dead flat, and you need a steady hand to keep the tool at this consistent angle.

As you hone, you may see a wire edge develop. You're always pressing the tool against the stone to keep it cutting, but right down at the very edge, the steel can get so thin that it bends away from the stone rather than being cut. You can take off this wire edge by bending it back and forth, stropping it on leather or your palm with alternating strokes on each side of the tool, always away from the cutting edge. Breaking off the wire edge leaves behind a microscopically rough, work-hardened edge that then needs further honing with a stone, which in turn leaves a finer wire edge which again needs to be stropped off.

I have become a convert to removing the wire edge by honing the tool on my finest stone alternately on the flat side and the bevel side—sharpening off the wire edge rather than breaking it off. I do wonder though, if the magic effect of new sharpening techniques comes from renewed attention to the routine more than anything else.

The proof of the edge is in the cutting—over time. A too-acute edge won't hold up, and microscopic scratches from insufficient polishing will leave tiny points that bend over and blunt the edge. Testing the edge on your thumbnail will tell you if it is sharp, but only wood and work will tell if the edge will last.

3 Hewer

Then from the olive every broad-leaved bough
I lopp'd away; then fell'd the tree; and then
Went over it both with my axe and plane.
Both govern'd by my line. And then I hewed
My curious bedstead out; in which I show'd
Work of no common hand.
   —*Homer,* The Odyssey

Hewing is classic wedge and edge work. You notch in every foot or two along the length of the log, then come back and split off the wood between the notches. Sometimes these chunks, called juggles or joggles, get split in half and are used to chink the spaces between the logs in an outbuilding, but, more often, they quickly disappear into the fire. Lighter scoring follows and then a cross-grain sweep of the broad axe. This last step can erase all evidence of the steps that went before it, but when enough of the story remains, it makes interesting reading.

We usually judge an artisan's skills by how well the finish of the product conceals the process used to create it. Tool marks on a piece are taken as the sign of unfinished or careless work. The hewer of wood, however, can finish the day proudly leaving behind a blow-by-blow testimony of the process on every timber. If you study the axe marks on the posts and beams of an old barn, you can almost see the person on the other end of the axe handle. You can see not just the strength, endurance, and control of the hewer, but also his judgment. The spacing of the notches (if traces remain) tells you how much wood the hewer figured he could split away. Looking at that same length of wood—considering the species and the run of the grain—would you have judged the same?

Spike Dog

For the sake of your axe and your back, you need the log elevated off the ground for hewing. At the least, make V-cradles in a couple of cross logs and roll the log up onto them. If you're hewing American-style, using a regular felling axe to make the cross-grain notches, you'll be standing on the log and need only enough elevation to keep the axe head clear of the dirt. If you're using German-style notching axes, then your log needs to be up on stout horses about waist high.

In either case, you need to keep the log from moving around as you work. If the log has any sweep to it, let that hang to the bottom. This is the most stable position, and you want this sweep captured in the wider dimension of the finished timber, rather than running in and out of the sides. Also helpful are spike dogs, giant staples driven between the cross logs and your timber. The ends of the dogs can be spike pointed or chisel pointed, with one end at right angles to the other to match the grain of the log and cross log.

Plumb Bob and Snapline

Even if you're not going to measure the timber precisely with a square and rule, you still want to be sure that the hewn faces are flat, not twisting or curving. Plumb bobs or winding sticks can take care of the twist; snaplines take care of the curve.

Winding sticks are just any two long, straight edges. You could use two framing squares or two yardstick-sized slats. With one set against the far end of the log, you can sight over one placed at the near end. Turning one or the

The spike dog holds the timber steady.

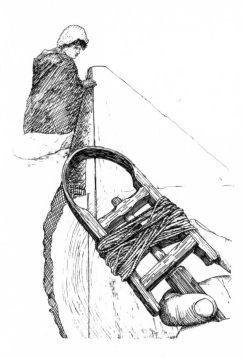

Snap lines down the length of the log to connect the plumb lines on the ends.

other until they look aligned, you can be sure that lines traced from them onto the ends of the log define the same plane.

Winding sticks work most conveniently in a horizontal plane. With logs, there's enough walking back and forth and fooling around to align winding sticks to declare them a nuisance compared to a plumb bob. With a plumb bob, you can drop a vertical line down one end of the log, walk to the other end and make another line, and be assured that the lines are in perfect alignment.

Now, with the same method used by Odysseus when making his curious bedstead, snap a chalked string to connect the vertical lines on the ends of the log with another perfectly straight line. A washer on the end of the string allows you to anchor it on your scratch awl, but a quick cut with the axe into the corner of the log will catch a knot tied in the end just as well. Unwind the line, inking or chalking or rubbing with willow charcoal as you go.

Always lift the snapline in the same plane that you are defining. On a perfect cylinder or perfect plane surface, you could snap any way you want, but the irregularity in a tree will cause the line to waver. To define a vertical surface, lift the string vertically. Your first snap may be just a guide for your drawknife as you remove a strip of bark to get a clear line on the second snap. The snapline is still the best way to get accurate lines on a piece longer than your two-foot square. Timber the size of a workbench top can't be laid out without it.

Chop notches across the grain . . .

. . . and then split off the chunks in between

Now all you have to do is remove all the wood on the outside of the line. Notching and splitting is the way. We know that the planes of weakness run down the length of a log, so, ideally, we could just swing with our axe and split all excess wood off the side with one blow. Even if we could get the split to run all the way to the other end of the log, we'd still be subject to any winding twist in the grain. We want the grain of the wood as our partner with us in determining the final shape—but only for about a foot or two. When the split reaches our cross-grain stopping cut—it's over. The waste piece falls away and we make the next split down to the next notch, and so on down the log. Hewing is systematic shaping with an axe—imposing culture on nature—with nature still as our half-partner in the process.

With a felling axe, stand on the log and swing down at the sides, notching just as if you were bucking the log, stopping at the chalk line, of course. You can probably sight down each notch and chop it plumb, but see that the log has not shifted out of plumb as you work.

You'll quickly learn to judge how much wood you can split away. Look ahead, though, for any upcoming knots in the sides. You won't be able to split off a chunk if a knot is pegging the wood to the tree. You need to notch directly in on big knots, adjusting your notch spacing accordingly. In a finished timber, this can make it look like the hewer had something against knots and was trying to chop them out.

German hewers often use a different method, notching with narrow axes, chopping vertically. The short-handled axe heads are perhaps eight inches long and two inches wide in the bit. Working as a team, two carpenters stand facing the elevated log and take quick turns chopping their side of the notch with downward strikes across the grain. Instead of splitting the chunks off with a swing at the end grain, they sink the bits of their narrow axes into the side grain of the chunk, just outside the chalk line. Pulling sideways with the short axe handle makes its head work like a froe, splitting off the chunk.

Broad Axe

The broad axe spoke, withouten miss,
He said the plane my brother is.
We two shall cleanse and make full plain,
That no man shall us gainsay.
    —Debate of the Carpenter's Tools, *1500*

Splitting removes 95 percent of the wood. The broad axe does the rest. Despite its size, the broad axe is the finishing tool, brother of the plane, used for slicing off the last quarter inch or so of wood when hewing a flat surface. From the early nineteenth century onward, the American broad axe blade has usually been beveled only on the side away from the timber, allowing the flat side of the blade to sweep down across the grain. The handle curves out away from the flat side, allowing your hands to keep clear of the timber surface.

Symmetrically headed but single-beveled axes can take the helve in either end of the eye and become right- or left-handed (determined by the forward hand that guides the stroke). Continental broad axes have elongated eyes forged for a lifetime of either right- or left-handed work.

Tool marks show that earlier American work was undertaken with double-beveled broad axes. Here, the bevel against the timber requires the axe to cant away from the line of the swing. This keeps the handle (and your hands) clear of the wood in either right- or left-handed work. Thoreau worked this way at Walden Pond: "So I went on for some days cutting and hewing timbers, also studs and rafters, all with my narrow axe."

The narrow axe still has work to do in any case, preparing the way for the broad axe. Standing on the log, swing down and make a line of vertical cuts, scoring across the grain, just up to the chalk line. Spaced three inches apart all the way down the length of the timber, these vertical score lines will break the power of the grain to draw the broad axe deeper than it should go. Any split that starts can go only three inches before it meets a score line. The grain is on a tightening leash.

Broad axes, timbers, and temperaments are too different for much more specific direction. You'll have more clearance for your knuckles if you work forward, moving from finished into unfinished areas. Apart from that, just work down across the grain with a sharp axe and don't hit your foot. Hewers of timbers that will be exposed in a fine building will work more carefully than those whose timbers will be hidden or be seen only by vermin, architectural historians, and the like.

Finally, I have always four-squared timbers by hewing down one side of a timber, then hewing the opposite face, then turning the log and finishing off the other two faces. Others, I have observed, see logic to hewing one face, turning that face to the top, then hewing the next two sides and then the bottom. Should you see someone hewing in a way that seems strange to you, think twice about correcting him. Anyone willing to hew timbers on a hot day is pretty much entitled to proceed in his own damn way.

4 Log Builder

*Most of the Houses in this Part of the Country are Log-Houses,
covered with Pine or Cypress Shingles, three feet long, and one
broad. They are hung on Laths with Pegs, and their doors too turn
upon Wooden Hinges, and have wooden Locks to Secure them,
so the Building is finisht without Nails or other Iron-work.*
   — *William Byrd,* History of the Dividing Line betwixt
      Virginia and North Carolina, *1728*

You'd think the first thing the English colonists would have done when they got off the boat at Jamestown in 1607 would be to drop some long, straight pines and lay up log cabins—quick and secure defense against the weather and unhappy locals. But what did they build? Timber-frame houses, just like they built back home in England. Just as the colonists had a toolbox in their hands, they also had a toolbox in their heads. As obvious as log houses seem to us now, they were not in the English mental toolbox.

Log building is simply not British. Have you ever seen a picture of the log cabin where Shakespeare was born? The idea of log building arrived in North America with settlers from the Baltic and Central Europe. It fit the environment so well that it quickly became part of the American identity. And if any of those original English colonists lived long enough to see that first log cabin go up, they must have dope-slapped themselves silly.

Log construction is essentially a stack of scribed lap joints. It developed to marvelous sophistication in Europe, but in America it reverted to the essentials. There still remains an impressive degree of regional variation in American log building, reflecting different settlement and cultural diffusion patterns. The common corner joints are saddle notching, V-notching, and dovetail notching.

Saddle Notches

Unhewn logs are most simply joined with saddle notches. Rough and ready, these are simply hollows chopped to cup over the ends of the round logs below. Each hollow, however is a scribed, custom fit, its depth determining the closeness of the logs.

Roll a log into place on the wall and there will be a gap between this log and the one directly beneath it. Notching will drop the upper log to close this gap, as well as locking the corners together. The question is, how much of the gap do you want to close?

When the notches are deep enough to fully close the gaps, they take away half of the end of each log. Upscale European log building closes these gaps completely—but that is not the cowboy way. American log building usually leaves some of the gap and makes abundant use of chinking and daubing to seal it—chinking of wood and stone and daubing of clay and mortar.

So, looking at the gap between the logs, set a stout pair of dividers to the distance you want the upper log to drop. Hold this setting on the dividers as you ride them over the tops of the lower logs, transferring the drop distance up onto the upper log. Throughout the scribing, you need to hold the dividers vertically. You're not trying to enlarge the profile of the lower log, just to transfer it directly upward.

When the contours are transferred, roll the upper log back and secure it for chopping. You can shape the hollow entirely with an axe, but a saw cut across the grain to the bottom of the hollow will make the chopping go faster. The hollowing is bevel-down work, and a side hatchet will give you more clearance on tighter curves. You'll have to climb over to the other side of the log to cut the opposite contour.

Scribe the profile of the lower log onto the upper one . . .

. . . then roll the upper log back and chop out the saddle.

If cut flush, saddle notches become too fragile, so the logs need to extend beyond the joints in the rustic, log cabin manner. Saddle notches work with round, unhewn logs, but if you want your building to last more than a year, you need to deal with the bark. Nature has spent millions of years perfecting its recycling system, and unless you take deliberate steps, it will treat your building just like a pile of fallen logs in the forest. When you fell the tree, it loses its active defenses against insects and fungi. Bugs smell the fallen tree and bore through the bark to lay eggs and feed. They carry in fungal spores that begin to grow — and the rest is humus.

Bark is the issue. Bark is waterproof. If it weren't, the tree would dry out and die. Nature's recycling system assumes that a fallen tree is going to retain its bark and stay moist inside. Peeling or hewing a log removes the waterproof bark, allowing the wood to dry before the bugs and fungi can go to work.

In the most funky circumstances, such as winter quarters for behind-the-lines partisans, you can lay up a cabin with the bark on the logs, wait until the bugs loosen the bark and then pull it off — stopping the natural process. Tobacco barns were commonly laid up from small pine logs with the bark still on them. Once the walls were up, the builders took a felling axe and chopped a three-inch-wide strip down both sides of each log. This allowed the wood to dry — and as long as wood stays dry and free from bugs, it can last forever.

V-Notches

V-notching will work in round logs, but it is most often used with logs hewn flat on at least two sides. Get the log up on the building where it is to go and mark its thickness on the two logs that it rests on. Roll the log back out of the way and cut the V-points on both of the bottom logs. Because the notches are custom fitted to the points, you can cut the V-points by eye with a hand axe. Two facets of the V-point make an angle with one another, from 90 to 120 degrees. Alternate your chopping on the shoulders and the facets, finishing off by pushing the flat of your hand axe across the facets like a broad chisel.

Now roll the top log back into place. See how far it needs to sink by checking the gap between the logs. You can scribe this gap upward from the lower facets with dividers, but if the facets are symmetrical and equally inclined, a framing square or a board with parallel sides works better. Hold the parallel guide on a facet and draw a line along the guide's opposite edge. Repeat this on the other facet, and the two lines will intersect directly above the apex of the V-point. Add more parallel lines to this upside down V until you reach the depth that you want the upper log to sink.

Repeat this upward scribing on the back face of the log and on the far end, and you have the outline of the notches. Roll the log over and go at it with your axe. When you have chopped down to the line, roll the log back into place and see how it fits. Make any adjustments and move on to the next log.

Diamond notching is a variant of the V-notch used in both hewn and round logs. Just like baseball diamonds, the ends are square but turned at 45 degrees to our perspective. Hew just the ends of the logs square, before laying up the

V-notching.

Trim flush with a timber saw.

A single slope set in alternating directions adds up to half-dovetail notching.

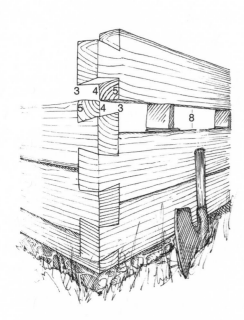

The oblique slopes of full dovetail notching require more precise layout.

walls, keeping the faces in the same plane at both ends. The top of the diamond forms the point of the V. You lay out and chop the V-socket in the undersides of the upper log just as with the standard V-notch.

Dovetail Notching

Half-dovetail notches are even simpler to cut than V-notches. The process is the same, except that you cut just one slope on the logs instead of two. The half-dovetail notches always slope down across the grain of the lower log to the outside. As before, scribe upward with parallel lines.

As with a cutting edge, you try to find a compromise angle for the slopes of the dovetails. The steeper the slope, the greater the interlock, but the more fragile the notches become. This is a good excuse to go study the log buildings in your part of the forest. Eventually, even the most deeply ingrained cultural patterns adapt to the strength of the local timbers.

V-notched and half-dovetail joints lock at the tops of the logs, but they can kick outward at the bottom. Full dovetail notches lock in the bottoms of the logs as well. To keep the notches strong as you chop them, V-notched and half-dovetailed logs can be left long and cut flush after the walls are laid up. Full-dovetailed corners are easier to lay out (and explain) if the logs are already cut to length.

I say easier—full dovetails can tax your powers of visualization, so we'll enlist some left brain arithmetic to help.

To completely close the gap between logs, the notching needs to remove half of each log's height. If the height of the log is divided into 16 parts, then half of that is 8 parts. If we made plain, flat, square notches, taking 4 parts from the top and 4 parts from the bottom of each log, that would add up to 8 parts—half the height of the log. Such square notching would close the gaps between the logs, but it would have no interlock at all.

Dovetail notching interlocks the logs, often using a 3-4-5 rule to still add up to 8 parts—closing the gap and creating the interlocking angles at the corners. Abstracted, the corners work like this:

3 4 5   parts up from the bottom of the upper log, over
5 4 3   parts down from the top of the lower log, makes
8 8 8

I have no evidence to prove a connection, but a framing square, with its standard 2-inch blade and 1 1/2-inch tongue is perfectly proportioned to lay out dovetail corners on 8-inch-high logs. Dividing an 8-inch-high log into 16 parts gives you 1/2 inch. The 1 1/2-inch tongue gives you three parts. The 2-inch blade gives you 4 parts (the width of the blade). The 5 parts is tricky. Starting at a point indicating 4 parts, drop back 3 parts (the width of the tongue). This gets you to 1 part. From there, go up 4 parts (the width of the blade again) and you've got 5.

With other log heights, you can make a template. This is simply a board, twice as long as the logs are thick. Make it 3 parts of the log height wide at one end and 5 parts wide at the other end. This makes the midpoint 4 parts wide.

You need precisely hewn and regular timbers for full dovetails. The timbers have lost any resemblance to logs and the structure is anything but a cabin. Moravian settlers brought this disciplined manner of building into North Carolina, and it certainly must have impressed the English.

5 Sawyer

The cause of the intemperance of the sawyers, say my informants,
was their extreemly hard labour, and the thurst produced by their
great exertion.
  —*Henry Mayhew,* London Labour and the London Poor, *1861*

If you know anyone named Sawyer, this is where his or her name comes from. Not from crosscutting, which everyone did, but from the specialty trade of ripping down the length of a log. English sawyers had a great reputation as drinkers, but Henry Mayhew ranked them only at thirty-fourth in his study of drunkenness in the trades, about equal with carpenters and wood turners—far below eighth-ranked French polishers, and complete slackers compared to the number one drunks of London, the button makers.

In England, sober, job-protecting sawyers pushed legislation through Parliament to ban the construction of horse-, wind-, or water-powered sawmills. The drinking sawyers did their part by burning down any sawmills that did get built.

In the colonies, there was anything but a surplus of labor, and a great demand for sawn timber. Water-powered sawmills were established almost immediately after settlement. Pit sawing held on longer in the South, where the wide coastal plain made water mill construction more troublesome than in the North. In the slaveholding South, much capital was tied up in owning labor. With little left for investment in industry, pit sawing continued as winter work.

Large estates often had their own sawpits. Thomas Jefferson had one at Monticello. In *The Merry Wives of Windsor* six children disguised as hobgoblins scare Falstaff when they "from forth a sawpit rush at once." Of course, many sawyers do not use a pit at all but rely on trestles and props to lift the log eight feet in the air. Trestles are certainly more portable.

Rip Teeth

Think of wood as a bundle of fibers all glued together, because that is what it is. Now consider how you would go about neatly severing these fibers to make slots first across their length (crosscutting) and then along their length (ripping).

We've already noted how, in crosscutting, the saw acts like a row of knives to slice across the fibers on both sides of the kerf. The leading edge of each tooth slopes back to slice the outer surface before the deeper part of the cut is made, just as you would instinctively angle a pocketknife to make a similar cut.

But this won't work along the length of the fibers. Instead of slicing them, knifelike teeth would ride along, and on either side of, the fibers. To cut this bundle along their length, you need to chop at them with a series of chisels, not slash at them with knives. A ripsaw is simply this, a series of chisels oriented so that they continually chop off the ends of the fibers as you work. Crosscut teeth are like knives; rip teeth are like chisels.

The rip teeth on a pit saw are one hundred 1/16-inch-wide chisels. Mount the saw on a bench with the teeth upright at about elbow height. Joint the teeth by drawing a six-inch file down the length of the saw until the tips of all the teeth are brightened.

Use a rounded file to bring all the gullets to a depth of at least three-quarters of an inch. The gullets on a pit saw undercut the faces of the teeth to give them a slight hook. Excessive hook in the teeth makes the saw too hungry, causing

Rip teeth work like chisels.

it to chatter and jump; too little hook cuts too slowly. You need less hook and smaller teeth for very hard woods, more hook and larger teeth for soft.

Set the teeth by alternately bending or hammering them to opposite sides of the saw. Strive for the minimum set that will keep the saw moving freely. You can always add more set if you need to, up to the theoretical maximum set of one-third the thickness of the blade.

Finally, sharpen the face of each tooth square across and then file the back slope until the flats from the jointing almost disappear. Perhaps it's just the sawyers getting accustomed to the newly sharpened blade, but a pit saw seems to hit its stride only after a few feet of sawing has mellowed the edges.

Lining Out

Before going on the pit, logs for boards of a constant width usually get hewn on two sides to a thickness equal to the width required. A log for structural timbers gets hewn on all four sides to a multiple of the final dimensions, say, eight by nine inches to yield six three-by-four-inch studs. You may see evidence of this in the timbers of old buildings. One or two sides of the timber will have axe marks—the remaining sides show that they were pit sawn.

The tools and procedure for lining out logs for sawing are the same as those for hewing, with the addition of a pair of dividers for pacing off equal planks. First snap the guidelines on the bottom side of the log and then roll it over into place on the pit. Now, at each end of the log, "plumb up" with a weighted string from the lines on the bottom to find the points to begin and end the lines on top of the log.

If the log has any bow in it, see that the bow is in the vertical plane during sawing. Boards with the grain moving across their faces lie flatter than those that have it moving in and out. For feather-edged weatherboards, saw the timber into one-inch boards, then tilt the log so their diagonal line is plumb and re-saw each inch board into two tapered weatherboards. Leave all the boards connected at the back end and break them apart only when the whole log is completed.

Timber cuts far more easily while it is still fresh and green. An eight-inch-thick, fresh-cut yellow pine can be easily sawn at a foot a minute. The green wood means you need to allow not only for the loss in the kerf (sawdust) but for shrinkage and any cupping that may need to be planed out.

Sawing

Pit sawing involves two men, and, traditionally, the top sawyer is the senior of the two, owner and caretaker of the saw. The pull stroke of the other man, the pitman, does the actual cutting of the wood, using the pitman's weight to his advantage as he rocks back on one leg. The top man keeps the cut on course and must pull the saw back up with his arms and shoulders alone.

You start a cut by hooking one tooth about 1/8 inch in from the edge. The pitman makes one sure downward stroke, and the work begins. Until the

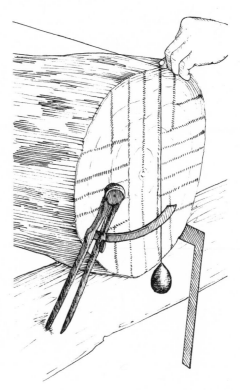

See that the cut is plumb before you start sawing.

When sawing feather-edged boards, tilt the log to keep the second series of cuts plumb.

The top sawyer stands close to the saw and steers with the tiller.

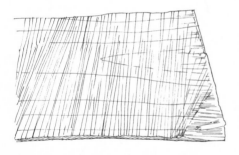

The marks on pit-sawn timber abruptly change their angle when the top sawyer steps back.

gullets of the teeth are buried in the kerf, sawdust will spray out to the sides. Once you've cut an inch into the log, though, the heavy sawdust falls directly down about two feet in front of the pitman. When the cut is far enough into the log to allow it, drive a wooden wedge into the kerf behind the saw. As the kerf progresses, drive in a second wedge further along, free the first wedge and leapfrog them along as you saw.

Just as with the two-man crosscut saw, the work is done on the pull stroke. To push would kink the saw; there is some pushing, but it must be the lesser force. The pitman bears off the face of the kerf during the upstroke with a gentle lift to the lower handle or box. Any resting of the pitman's arms on the upstroke will quickly wear out the top sawyer. The top sawyer can help his partner by pushing down just enough to help the cut along — but never so much as to cause the saw to chatter and jump.

The top man guides the saw with the long-handled tiller. Twisting the tiller steers the cut down the top line. Bending, or "throwing," the saw causes it to lever against the top edge of the kerf and forces the bottom in the opposite direction. The pitman can steer to some extent with his handle, the box, but the leverage from above is essential. All the steering in the world can be defeated by a cut that has fallen out of plumb. The log should be shimmed with wooden wedges and secured with spike dogs, but check the plumb if the saw starts wanting to run off line.

A good top sawyer stands close to the saw, nose to the teeth, keeping the saw cutting vertically. Every foot or so, the top sawyer has to step back to keep from sawing himself in half. This changes the angle of the cut, and this shows in the saw marks — the diagnostic pattern of pit-sawn timber.

So it goes, sawing and wedging the kerf until the pitman calls out to warn the top sawyer that they are about to cut into a support timber. If the log is to be cut into more than two pieces, all the kerfs will be sawn up to the first support before continuing any one kerf to the end of the log.

Once all these cuts are complete, you move the support forward and knock loose the bottom box, then pull out the saw and reinsert it on the near side of the support. This is where sawyers show panache. The top sawyer draws the blade from the kerf like a saber from a scabbard, walks with it along the log, and stabs it through the next kerf. Beneath the log the pitman holds the box poised to capture the end of the blade the instant it appears. The pitman drives the wedge tight in the box, and the sawing continues.

6 Frame Carpenter

The house-builder at work in cities or anywhere,
The preparatory jointing, squaring, sawing, morticing,
The hoist-up of beams, the push of them in their places,
    laying them regular,
Setting the studs by their tenons in the mortices according
    as they were prepared,
The blows of mallets and hammers, the attitudes of the men,
    their curv'd limbs,
Bending, standing, astride the beams, driving in pins,
    holding on by posts and braces . . .
    —Walt Whitman, "Song of the Broad-Axe," 1891

Cut the lower half of the lapped notch in the wall plate by sawing across the grain and then adzing out the waste.

Drop the joist into the notch and mark the width of the upper notch.

Pull the joist back out and gauge down from its top edge.

One wall at a time, one floor at a time, one roof truss after another — the carpenter will measure, cut, and drive timbers together. There will be sweat in every beam, skill in every cut, but from the first tree felled, the carpenter's intelligence has imagined the whole.

In the carpenter's mind, the long oak timbers stacked in the yard were sills even as they stood in the forest. The first glance at the growth rings in the first chips from felling the pines revealed their potential as beams. Narrow rings indicate slow growth and, in pines, the subsequent long fibers of strong, stiff wood.

Scribe and Square

Even the strongest stuff needs a carefully thought-out design and skillful execution to make a strong building. In the old way of carpentry, you scribe each joint by superimposing timbers on one another and marking the intersections as they happen to fall. This scribing copies the variations of one piece onto the other piece so you can cut them to a close fit. To lay out a wall, you arrange the sill and the top beam parallel to one another on the ground. Set the corner posts spanning them, properly spaced with the corners squared. With a scratch awl, scribe lines where they cross one another (they should be longer than needed). These lines become your guides to saw the shoulders of the tenons and to find the locations of the mortises. With the outer frame connected, subsequent timbers get custom fitted within it.

Few things get measured — but everything gets numbered, because each piece will fit in only one place. Nothing can be reversed; no brace can be exchanged for another. Come raising day, you depend on the Roman numeral "marriage marks" to get every piece back where it belongs.

Some time around 1815, a new method emerged. If guns and clocks could be made with interchangeable parts — why not buildings? Instead of the old scribe rule, carpenters began building by the square rule. With the square rule method, you can design a building on paper and start making braces and posts in your basement, knowing that they will fit together on raising day.

The square rule still uses the same rough-hewn timbers of the scribe rule, but measures and cuts the joints as if every piece were perfectly dimensioned. Say you've got a heavy hewn timber with the normal variations in dimension. Guided by chalk line and square, you sink and level the seat around every mortise in this timber to align with the perfect ten-by-ten hiding within it. Every brace tenon can be struck from the same template, because the ends of every timber are perfect. No matter how funky the lengths in between, the joints reside in a perfect world.

Lap Joints and Gauges

The simplest timber joints are lapped notches, just like those used in log buildings. For plain half-laps, you take away half of each timber at the area of overlap. You'll find half-laps at the very bottom of a building in the corners where the sills intersect, and in the very top where the rafters join at the peak.

The laps usually get pegged for security, but as always, the joint carries the load, not the peg.

Half-laps bring equally thick timbers into the same plane, but few timbers are equally thick. If one side of the assembly is more important than the other, this is termed the face side. You ensure that the face side is level by making all measurements from that surface and letting the back side take care of itself.

When using a gauge, the rule is to always gauge from the face side. The gauge is just a small beam with a scratcher set through an adjustable fence, used to mark a line or lines parallel to any surface. For a half-lap, set your gauge for half the thickness of your timbers and run it around with the fence riding on the face sides (say, the upper sides) of both timbers. When the lap is cut to those lines, the face sides of the timbers will be dead level, and any difference in thickness thrown to the back sides.

Saw the shoulders of the notch, . . .
. . . split out most of the waste, . . .

The face side is not just a matter of appearance. For example, if joists are to receive flooring, then their tops need to be all on the same level. This makes the joist tops the face sides—the side you measure from, the side where the fence of the gauge must ride. If the notches in the top of the plate (the top beam of the wall) have already been cut evenly to the depth of a chalk line, then gauging down from the tops of the joists will ensure that they will sit level with one another. The square or any straightedge will transfer the width of the lower timber onto the upper one, and the gauge gives you the depth of the notch.

The saw defines the shoulders of the joint, making stopping cuts for the splitting to follow. A four-inch-thick, half-seasoned timber makes slow going for a fine-toothed saw. Undignified as it seems, the same bucksaw used for cutting firewood can quickly cut the shoulders of a timber joint. In seasoned, hard timber, however, a coarse saw will jump around, so the teeth have to match the task.

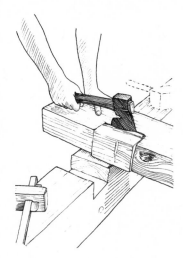

Splitting saves time only if it doesn't weaken the joint or ruin the timber. You can safely split away perhaps two-thirds of a lap joint with a hatchet—more if the grain looks good. The risk is a split that turns deep into the timber, guided by a buried knot. Work your way deeper as you repeatedly turn the timber to work from the other side. After the first big chunk comes out, the smaller waste wood will bend away from the hatchet blade. This bending waste will exert less leverage, and the hatchet's edge will do more work than its wedge.

Of course the other risk of splitting out a joint with a hatchet is striking the wrong spot. Many folks use their hatchet as a broad chisel, setting the blade on the wood and striking the back with a mallet.

Framing Chisels, Slicks, and the Besaiguë

A framing chisel is the proper tool for this job. Big and strong, the wooden handle in the heavy socket preserves both the mallet and the tool. When the splitting is done, hand pressure alone can guide the framing chisel's edge to shave the face grain smooth. With taps of the mallet, it can shear off the last sixteenth of an inch of end grain to bring in a shoulder. Slid down a split billet, it can shave it into a rounded peg.

. . . and finish with the framing chisel.

Gauging from the top edge assures that the tops of the joists will be even.

1. – CARON Juliette, née le 6 Mai 1882, à Senlis (Oise)
La seule femme en France exerçant le métier de charpentier
Travaillant actuellement aux casernes de Montluçon

Fin de siècle carte postale *of Juliette Caron, a French carpenter, with her besaiguë.*

The framing chisel can do all of this edge work only if it is sharp. It's a single bevel tool, ground and honed to 30 degrees. As with any edge tool, no matter how finely you hone it, the blade has a microscopic sawlike roughness. If you slide the chisel to the side as you push it, the saw effect works in your favor.

Along with swinging an axe and driving a chisel, there is a third way to strike—exploiting the inertia of a heavy blade jabbed in short strokes. A framing chisel or a sharp side axe held flat against the cheek of a tenon will shear it smooth if it is repeatedly slid quickly forward. Direct pressure on the tool creates a small starting surface that you can easily expand by using it as a slide for the jabs that follow. Two tools designed specifically for the inertial jab cut are the slick and the French besaiguë.

Slicks look like giant chisels with a wooden handle long enough for a two-handed grip. Boat builders use them for fairing the constantly changing bevel of planking. They are impressive and powerful tools—popular among American timber-frame carpenters.

The besaiguë (pronounced *bez-ai-gue*, as if someone tried to strangle you before you could get out the last syllable) is the double-sharp French carpenter's tool. The ends of the four-foot-long iron body have chisel blades set at right angles to one another and a hollow iron handle at midlength. Working with either the broad chisel end or the mortising chisel end, your left hand positions the blade above the intended cut. With the other end of the tool resting on your shoulder, your right hand is free to make the jab. With the handle down, the flat of the broad chisel is to the left. If you need the flat to the right, you turn the besaiguë on its long axis and grasp the handle upside down.

The narrow mortising chisel on the other end of the besaiguë works not only with the inertia of the tool, but with the leverage of the four-foot-long iron body. It can easily rip out the wood between auger holes for shaping a mortise, another lever-edge.

Sawhorse and Square

The besaiguë works to its best advantage when the timbers are lying low to the ground. It's a French thing, but like most other carpenters, those in France also elevate their work on the horse foaled of an acorn. Sawhorses for heavy oak timbers are more like benches made from a log; flattened on top, with three or four legs set into holes bored through it.

For stability and strength, the legs of a sawhorse splay out to the sides and to the ends. When making a heavy horse with augered leg holes, you can just eyeball the splay of the first leg, drive it in and then use it as a guide to bore the remaining holes with equal splay. In a light sawhorse, made from sawn stock and held with screws and nails, you work by more calculated layout. Finding these leg splay angles gives us a brief lesson in using the framing square. Don't run away.

The square is fixed at 90 degrees and marked in inches, counting out from the heel. This and an angle table are all you need to measure and mark any angle. For example, the rafter feet of a half-pitch roof must be cut at 45 degrees. To find 45 degrees across a timber, position the square on the timber's edge so that the square crosses at any equal number on both limbs.

A light sawhorse runs about four feet long and two feet high. The splayed-out, 1-by-3 legs lap into angled seats cut across the 2-by-4 top beam. Twenty-two degrees is a good endwise splay for the legs. To find 22 degrees, set the square on the top beam of the horse-to-be so that it crosses at the 8 inch mark on one limb and the 20 inch mark on the other. Draw this line across the beam of the horse and you have the end splay.

Another way to express an angle is as a ratio. When we get to dovetail angles, we'll refer to an angle of one in six—a one-inch rise over a six-inch run. On the sawhorse, the legs also need to splay out to the sides enough to make the feet sit at least 14 inches apart—or each leg seven inches out of plumb. Since the horse is to stand 24 inches high, the ratio is 7 in 24. Set a straightedge to cross the square at these respective inch marks and you have the second angle for the notch in the top beam.

Now you have the angle, but how do you transfer it to the narrow top beam of the sawhorse? Simply derive smaller units from the square instead of full inches. The 7 in 24 ratio remains the same if the units are feet, inches, eighths, or sixteenths. If the angled notch runs 7 units in on the top of the beam and 24 units down on the side, the leg fitted into it will splay accordingly.

Angles from the Square

To lay out angles between 1 and 45 degrees with the framing square, set it to cross the base line at the numbers indicated.

1°	3/8	20		24°	8 1/8	18 1/4
2°	11/16	20		25°	8 7/16	18 1/8
3°	1 1/16	20		26°	8 3/4	18
4°	1 3/8	19 15/16		27°	9 1/16	17 13/16
5°	1 3/4	19 15/16		28°	9 3/8	17 11/16
6°	2 1/16	19 7/8		29°	9 11/16	17 1/2
7°	2 7/16	19 7/8		30°	10	17 5/16
8°	2 3/4	19 13/16		31°	10 1/4	17 1/8
9°	3 1/8	19 3/4		32°	10 5/8	16 15/16
10°	3 1/2	19 11/16		33°	10 7/8	16 3/4
11°	3 13/16	19 5/8		34°	11 3/16	16 9/16
12°	4 3/16	19 9/16		35°	11 1/2	16 3/8
13°	4 1/2	19 1/2		36°	11 3/4	16 3/16
14°	4 13/16	19 7/16		37°	12 1/16	16
15°	5 3/16	19 5/16		38°	12 5/16	15 3/4
16°	5 1/2	19 1/4		39°	12 9/16	15 9/16
17°	5 7/8	19 1/8		40°	12 7/8	15 5/16
18°	6 3/16	19		41°	13 1/8	15 1/16
19°	6 1/2	18 15/16		42°	13 3/8	14 7/8
20°	6 13/16	18 13/16		43°	13 5/8	14 5/8
21°	7 3/16	18 11/16		44°	13 7/8	14 3/8
22°	7 1/2	18 9/16		45°	14	14
23°	7 13/16	18 3/8				

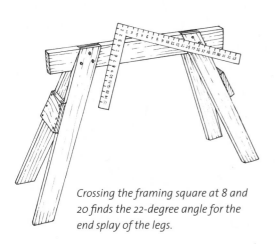

Crossing the framing square at 8 and 20 finds the 22-degree angle for the end splay of the legs.

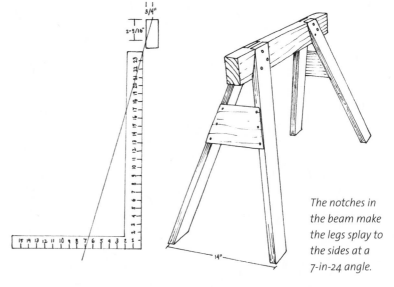

The notches in the beam make the legs splay to the sides at a 7-in-24 angle.

The miter lap joint allows you to fit a stud into an assembled frame.

The dovetailed lap joint can also be fitted into an assembled frame, but is strong in tension as well as in compression.

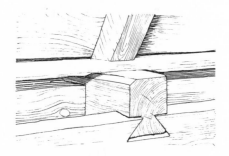

The dovetail tenon connects a stub joist to the plate.

You can also, and more handily, start with a known dimension and find the other one. Say you know you want the legs to sit 3/4 inch deep into the top of the beam. With the straightedge crossing the square at the 7 and 24 as before, measure in to find the point where the straightedge is 3/4 inch from the edge of the square. The other leg of the right triangle thus formed is the distance to mark down the side of the beam—in this case, very close to 2 9/16 inches.

With all this Cartesian geometry, you'll be glad to get to sawing the diagonals and chiseling out the waste in the top beam notches. Since the legs splay out, you'll need to cut them about 25 1/2 inches long to have a 24-inch-high horse. The tops and bottoms of the legs must be cut with a compound bevel. If mathematical precision is your priority, you could figure all this out with advanced geometry, but that would be putting Descartes before the horse. Instead, just stick the leg stock into one of the notches in the top beam and saw it off flush with the top. Turn the piece end for end, set it in the opposing notch extended to the desired length, and saw it off flush with the top. This leaves you with one finished sawhorse leg and the start of the next one.

Miter Lap

A few more laps. The sloping seat in the sawhorse beam causes the straight leg to kick out. But fit the reciprocally sloped end of a timber into it and you'll have a miter lap joint. This old joint allows you to fit studs into a wall, even after the main timbers have been framed and pegged. The stud gets marked where it overlaps the beam and trimmed to a shoulder and beveled tongue. Fit it against the beam again and mark around the overlap. Saw the diagonal and chisel out the wood in between so that the lap can drop in flush with the surface. It's not a great joint, because it can get pushed out against the single spike that holds it in place, but colonial Virginia houses are full of them.

Dovetail Lap

Rectangular frames can rack into parallelograms unless diagonal braces add the immutable strength of triangles to the mix. Braces in the corners strengthen the building against wind load by blocking the closure of the right angles. Triangles in roof trusses put some timbers into tension while bracing the compressed timbers against bending. Braces are critical, but can be only as strong as the joints that connect them.

Both the miter lap and the half-lap can pull free, but the dovetail lap can come out only the way it went in—at right angles to the strain. The dovetail keeps the joint strong in tension, and the shoulder resists compression. Because it is a lap joint, it can be let into a frame after it has been assembled.

The dovetail lap becomes a half-dovetail when it connects timbers diagonally, as in braces and collar beams. In right-angled tying joints, the dovetail remains full. The top of a flared "gunstock" post carries both the plate and the tie beam on tenoned shoulders. The underside of the plate is partially cut away to form the dovetail that drops in the corresponding socket in the top of the plate.

Dovetail Tenon and Bridle Joint

These two joints are halfway between lap joints and mortise and tenon joints. The sliding dovetail tenon has an Escheresque ambiguity—which is the tenon and which is the mortise? Both parts of the joint have positive and negative spaces. In any case, the dovetail tenon often serves to connect a stub joist to the plate where a stairwell or chimney interrupts the pattern of floor joists. It's simple enough to cut with a chisel and saw but has minimal rigidity unless precisely fitted into deep, hard wood.

A bridle joint is an open mortise and tenon—quick to cut because you can saw the inside cheeks of the joint and remove the root with an auger. In carpentry, the bridle joint usually joins the ends of timbers. In other trades, it gets more mid-timber action. The legs of a chair often connect to rockers with a bridle joint.

Laying out and cutting the bridle joint introduces the advantage of using tools in matching incremental sizes. If your gauge, your auger, and your chisel all cut an inch-wide swath, then you're well equipped to cut one-inch-wide mortises. Other sizes will be more bothersome without tools to match them.

The bridle joint at the peak of the rafters is an open mortise and tenon.

Mortise and Tenon

Joseph Moxon in his 1678 *Mechanick Exercises* struggled with proportioning the basic mortise and tenon joint. He observed that "if one be weaker than the other, the weakest will give way to the strongest when an equal Violence is offer'd to both. Therefore you may see a necessity of equallizing the strength of one to the other, as near as you can. But because no rule is extant to do it by, nor can (for many Considerations, I think,) be made, therefore this equallizing of strength must be referred to the Judgement of the Operator."

Proportion is more critical than size in mortise and tenon joints, for enlarging one element means reducing the other. Tenons can be quite small if they are just holding a timber in place, but in equally sized timbers, I follow the old rule of thumb, making the width of the mortise and tenon more than a third, but less than half of the timber's thickness. For 2-inch-thick timbers, this would be 7/8 inch, but it you have only a 3/4-inch or 1-inch chisel, that's probably the width you'll use.

A framing chisel is stout enough to work as a wedge . . .

. . . and sharp enough to cut with its edge . . .

The mortise and tenon joint is founded on the width of the chisel, because all the work of mortising is repeated cross-grain chopping, taking out chips until the mortise reaches its final size. For smaller joints, there are specific mortising chisels, but in carpentry, it's another job for the framing chisel.

The chisel itself can serve as the layout guide. If the sides are parallel, you can align it on the timber and trace down the sides to indicate the width of the mortise. Since the mortise and tenon must be the same size, the chisel serves as the template on both pieces.

The chisel can give you the width, but not the placement, of the mortise and tenon. The framing square serves well, and if the timbers are smooth enough, so does a gauge with two teeth set to the width of the chisel. Gauges with

. . . when paring the cheeks of a tenon.

adjustable teeth and fences are more common in joiner's work, while carpenters more often make a fixed gauge that won't lose its adjustment when stepped on, dropped, or thrown.

As always, gauge from the face side on both the mortise and the tenon pieces. Because a mortise can't easily be made smaller, or a tenon made bigger, I generally cut the tenon first, sawing the shoulders and splitting and shaving the cheeks. This done, I lay the tenon on the mortise lines of the other timber and mark its crossing points.

You can make a neater mortise if you start shallow and work deep. Starting perhaps 1/4 inch in from the ends of the mortise, set the framing chisel across the grain with the flat face to the end. Give the chisel a whack with the mallet to drive it in no more than 1/4 inch. Reset the chisel a little farther along and whack it again. The chip should split free and move into the space made by the previous cut. Continue this shallow cutting down the length of the mortise, stopping 1/4 inch from the other end.

Now you can start mortising in earnest. The process is called "chopping," an odd-sounding name until you think about felling trees with an axe—the cross-grain chopping action is the same. Some folks like to start in the middle and work outward. Others march back and forth. In either case, the bevel of the chisel faces the uncut wood as you advance. Each strike of the chisel goes a little deeper, because the preceding strike has made room for it.

The mortise needs to be a bit deeper than the tenon is long to allow for shrinkage. As you go deeper, the waste builds up and has to be dug out with the chisel levering against the ends of the mortise. Don't think you can't break a framing chisel when using it as a lever. Levering out the deep chips dents the shoulders of the quarter inch of wood left at the ends of the mortise—but that's why you left it there. You take out that last bit at the ends only when all the rest of the wood has been chopped and cleaned to full depth.

Now and then you may have to turn the blade with the grain to macerate the chips. This is a good time to contemplate the magic of mortising. If you were to drive the framing chisel hard enough, you could split the timber—but only if the edge were aligned with the grain, acting as a wedge pushing between the fibers. When mortising, chopping across the grain, the edge is dominant, severing a chip and pushing it aside. You can strike across the grain as hard as you wish, as often as you wish, and never split the timber.

Boring Machine

The first boring machine salesman must have had an easy job—until the competition got going. True, all boring machines work about the same as they bore the hole. Two hand cranks turn a bevel gear meshing with a second bevel gear on the shaft that holds the interchangeable auger bits. This lets you set the machine on the timber, sit on it, and crank away as the machine guides the auger with speed, power and precision. The gear frame slides down a rack as the lead screw of the auger pulls it into the wood.

Mortising is a matter of repeatedly chopping across the grain.

Boring machines can remove much of the waste.

When the hole is done, the boring machine needs to retract the auger. It's here, when the auger reaches the bottom of the hole, that American inventors went to work. You have to take a few backward turns to disengage the lead screw of the auger, but then the boring machine kicks in again. With the flip of a lever or the slip of a gear, your boring machine retracts the auger by climbing a spiral shaft, or by winding up a leather belt, or by dropping a rack into a pinion. The auger retracts, and you're ready for the next hole.

But wait, there's more! Some allow you to set a depth stop for automatic retraction. Some have a hollow drive shaft for long augers. This one has two heads for different speeds. That one changes angles with the turn of a screw.

When you have many holes to bore, a boring machine can make the job much easier. Anyone with moderate strength can use it. You can just set it over where you want a mortise, get it started, and watch as passersby take turns to try it out. Anyone can break a boring machine as well. The malleable cast iron parts might survive being knocked off a timber onto soft ground, but a cement floor can put your investment in pieces.

Corner Chisel

The carpenter's corner chisel squares a mortise after the auger roughs it in. It's an old idea, but for carpenters, the advantage was so marginal it hardly justified the cost. As tools became cheaper, specialty tools like the corner chisel became more common at building sites.

Carpenters, working in straight-grained timbers, can readily square corners with a framing chisel. The edge goes in across the grain and splits off the waste toward the auger hole. For wheelwrights, however, mortising elm hubs called for a more tightly angled corner chisel called the bruzz. Elm has interlocked grain and won't split away neatly like oak or pine. The best way to clear the corners of the tapering mortises in an elm hub is with a tightly angled corner chisel.

Corner chisels are often found sharpened to a point at the intersection. This make placement easier, but also makes the tool more spikelike. It doesn't take too many embarrassing stuck corner chisels to teach you to take smaller bites and work your way into the corner.

Drawboring

Drawboring uses slightly offset peg holes to force a mortise and tenon joint tighter together. First you bore a hole through the cheeks of the mortise, without the tenon in place. You then insert the tenon and mark the position of the hole on the tenon. You then pull the joint apart again and bore through the tenon, offsetting the hole about one fourth of its diameter toward the shoulder of the tenon.

When you reassemble the joint and drive a tapered iron hook pin through the offset holes, the tenon will be pulled tightly into the mortise. The hook pin compresses the wood and smoothes the way for the wooden pins to follow,

The framing chisel clears out the remainder . . .

. . . and the corner chisel finishes up.

Drawboring begins with the peg hole through the cheeks of the mortise.

LEFT: *The tenon goes into the mortise so it can be marked . . .*

MIDDLE: *. . . and is then withdrawn to let you bore the hole through the tenon, . . .*

RIGHT: *. . . offset by about a quarter of the peg diameter toward the tenon shoulder.*

and the hook is there to help you pull it back out. The deeper you drive the tapered hook pin into the hole, the more it will force the joint together—and the greater the risk of shearing out the end of the tenon.

Drawbored or not, pegs, pins, trunnels, whatever you call them, can run up to a quarter of the width of the tenon. Make the wooden pins from strong, dry, split stock, and leave them long and tapered so they can pull up the joint and be driven tighter as needed.

Beams and Tusk Tenons

If a beam is unable to support its own weight, you could just make it larger—but that would make it heavier as well. The trick is to make the timber larger in the right direction. The strength of a rectangular beam varies directly as to changes in its width, but as the square of changes in its depth. Make a timber twice as wide and it gets twice as heavy and twice as strong. Make a timber twice as deep, and it gets twice as heavy, but four times as strong.

How might you connect joists to a big summer beam running down the middle of a room without weakening it? The beam has to shoulder its load even after you cut it full of holes. The trick is to cut into the part of the beam that's doing the least work. Any beam is under compression on top, and under tension below. That's how an I-beam saves weight—the top is squeezed, the bottom is stretched, and you leave out the fat in the middle. In good timber design, mortises only cut into the waist of the beam, the neutral area, not into the top or bottom.

The tenon can fail as well. A tusk tenon gains more support from stepped reinforcements above or below the main tenon. Where the construction permits, the shallow shoulders of tusk tenons can gain extra security by having the narrow central tenon pass all the way through the mortised beam. This tongue extends out far enough for a tapered wedge or key to pass through and keep the joint pulled tight. These days, any through tenon secured by a

Tapered pegs passing through the misaligned holes draw the joint tight.

transverse wedge is called a tusk tenon. This gives you more opportunities to make the observation about timber framing in Alabama, where the Tusc-a-loosa, but in the old sense, the tusk was only the strengthening shoulder of a shallow tenon.

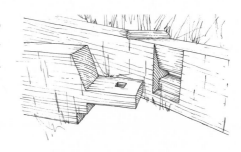

The tusk tenon is designed to maintain the strength of the mortised timber.

Trait de Jupiter

It's called a lightning bolt, but it took centuries. For generations, carpenters worked to solve the problem of the timber scarf—how to connect the ends of two timbers into a single long beam with a joint that will resist tension, compression, bending, and twisting. By the mid-thirteenth century, English carpenters almost had it, a lightning-bolt-shaped scarf with the ends undercut to interlock under pressure from the central transverse wedge. Through the rest of the century, they found ways to make it even stronger, but the improvements were marginal compared to the added cost. They had found the optimal solution—but did not know it at the time.

Thus flows the arc of all endeavor. Once solved, a problem is gradually forgotten, and with it goes the thinking that went into the solution. Further generations of carpenters knew how to make a scarf joint, but with the problem solved, they began to forget why. They kept tinkering, simplifying this element, elaborating that element, and producing scarf joints that were often inferior to their grandfathers. Only when the "improved" joints failed, while the old ones stood firm, did carpenters suspect that they had passed through one small apogee of their art.

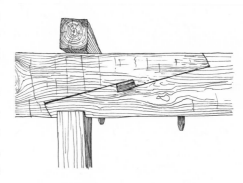

The trait de Jupiter scarf joint joins timbers end to end.

Spanish Windlass

The great mallet called the commander serves to drive the completed frame together. When you need more constant pressure, loop a rope between two timbers, insert a winding stick and you have a Spanish windlass. You can make it powerful enough to pull a timber frame together or light enough to tension a bow saw. A Spanish windlass makes a strong clamp to tighten chair legs as the glue sets—as well as an easily adjustable tie for bentwood bows for the chair back.

The term Spanish windlass just as commonly, and more appropriately, refers to a capstan improvised from two stout spars and a single rope. Say you want to pull a log up a slope. Tie one end of the rope to the log and the other end to a tree. Stand a spar at the midpoint of the rope. Make a tight loop in the rope, catch it with the other spar and start winding it around the lower part of the upright spar. Someone has to hold the upright spar vertical and see that the lines feed on at the same level while the other person walks the horizontal spar around.

Be careful. You can store tremendous energy in ropes and elevated or flexed timbers. Natural fiber ropes don't stretch as much as some synthetics. Should something give, a stretching rope lets go like a lethal rubber band—with you as the bug on the wall.

The Spanish windlass can help draw the frame together—and knock your teeth out if it slips.

Block and Tackle

Suppose you have a rope and a pulley, and you want to pull a log six feet across the ground to the base of a tree. Anchor the pulley to the base of the tree and run the line through it with one end tied to the log. To pull the log six feet, you have to pull the rope six feet. The mechanical advantage is zero and you might as well just drag the log with the rope and forget the pulley. All the pulley did is change the direction of the force.

Now, tie one end of the rope to the tree, tie the pulley to the log, and run the line through it. Thus arranged, to move the log six feet, you'll have to pull the rope twelve feet. In exchange for the double distance, you'll have double strength. One hundred pounds of pull can move two hundred pounds of load.

When you're skidding a timber, you can stand anywhere and pull. Lifting a timber, though, you probably want to stay on the ground, so you add one more pulley at the top of the gin pole to change the direction of the force. Now you have that 2:1 advantage again, and you can pull the rope with all your weight.

Rule of thumb: pull from the anchored end and your advantage equals the number of passing ropes; pull from the moving end and you subtract one, because the last pulley just changes the direction of the force.

This particular block and tackle arrangement gives a 2:1 mechanical advantage.

Crab and the Deadman

The crab, or capstan, also trades distance for strength. If the axle where the rope winds is eight inches thick, but the arms are eight feet long, the four people pushing it will have a 12:1 mechanical advantage.

The crab itself has to be solidly anchored. If there is no convenient tree, a deadman anchor will serve. The deadman is just a log buried in an undercut trench dug at right angles to the load—so what could go wrong?

Murphy's law works even as you sleep. Say you're raising a huge oak earthfast barn frame. The cable runs from the crab, under the frame to a tree, through a block, over a pair of shear legs and back to the frame. The blocks and lines are plenty strong, the shear legs well lashed, the snub lines secure. You've thought of everything, but a gentle rain the night before raising day softens the soil around your anchor in the earth. Halfway up with the oak frame, the deadman starts pulling free. Some shouting and some quickly pointed fence rails speared

The crab, anchored by a deadman, pulls the line through a pulley attached to a tree. The line, supported by the shear legs, then raises the massive oak frame.

into the walls of his grave keep the deadman down in his hole. The wall goes up—another close call, and a bad call at that.

It's not as if the battle of Gettysburg hinged on your charging forward with the raising. Wisdom should have commanded the wall to be backed down until you could dig a deeper pit. What might have happened if your improvised fence rail solution failed?

Shingles, Nails, and the Preacher

The heavy carpentry is done and the raising crew has gone home. Now comes work that overlaps the trades of the carpenter and the joiner. The wavy line between carpentry and joinery is traditionally defined by the use of planed stock. "Joiners work more curiously, and observe the Rules more exactly than Carpenters need do," says an old manual. But there is plenty of hammer and nail work in this boundary as well. Roofing and siding doesn't require the most advanced skills of carpentry or joinery. Still, it's not the worst work in the world, and, like anything, can be done well or poorly.

At any given point, a shingle roof is three layers deep. Each shingle shows less than one-third of its length to the weather, with the remaining two-thirds covered by the shingles above. With shingles 25 inches long, for example, 8 inches would be exposed to the weather, and the nailers would be spaced at 8-inch intervals.

Shingles swell and shrink as they are alternately soaked and parched by the weather. If you are nailing up freshly split shingles, fit them right up against each other, for they are as fat as they are going to get. Dry, seasoned shingles need to be spaced about a quarter of an inch apart. Any closer and when they get wet and swell, they will push against each other, buckle, and pull loose.

Some rain always gets in, but if the shingles are nailed to narrow one-by-three shingle laths, or "nailers," they can quickly dry with no harm done. One nail in each shingle will do, placed off center where it will be covered by the next course. This results in at least two nails through each shingle, as the nailing for the next course passes through the top ends of the shingles beneath.

With the roof on, the siding boards come next. If you are trimming the ends of plank siding flush to vertical corner boards, you have to get the fit just right. Too tight and they begin to spring out the corner board. Too loose and you have a gap for rain. The U-shaped preacher can help you get it exactly right. Scribe the line on the plank with the preacher pressed against the corner board.

Thin shingles and siding are easily split by the wedge effect of a large nail. Cut nails are square on their points and tend to punch through the wood, rather than wedge it apart as the tapered points of wire nails do. Properly oriented, cut nails are less likely to cause splits than a wire nail of the same size. Cut nails have two parallel sides and two converging sides, so orient them with the converging sides facing the ends of the board. Cut nails prevent splitting by directing the wedge effect perpendicular to the weakness in the wood—the same as when you chop a mortise with a chisel. Even the humble nail knows the strength of the grain.

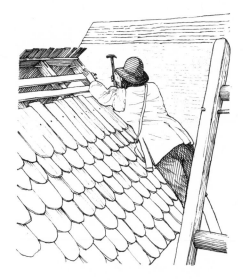

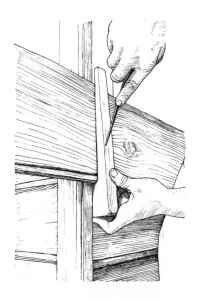

Nailed to the narrow laths, the shingles are well ventilated underneath.

The U-shaped preacher helps you fit weatherboards tight between the cornerboards.

Three Dumb Nail Tricks

In heavy timber carpentry, the dangerous work is finished by the time the nails come out. Perhaps that explains the considerable lore based on fooling around with nails. Here's what carpenters were up to in 1678, as reported by Joseph Moxon:

> A little trick that is sometimes used among some (that would be thought cunning Carpenters) privately to touch the Head of the Nail with a little Ear-wax, and then lay a Wager with a stranger to the Trick that he shall not drive that Nail up to the Head with so many blows. The stranger thinks that he shall assuredly win, but does assuredly lose; for the Hammer no sooner touches the Head of the Nail, but instead of entring the Wood it flies away, notwithstanding his utmost care in striking it down-right.

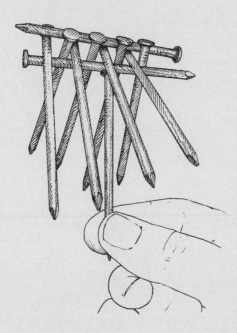

The ten-nail balance.

If you find yourself lacking in the earwax department, you might try betting that you can balance ten nails on top of one nail. The solution uses the rope walker's trick of lowering the center of gravity. First, lay one nail flat on the bench. Lay eight nails across this one, heads hooked over the first nail, points set in alternating directions. Lay one more nail between the row of heads. Lift the lower nail, and as the points of the crossed nails hang down, the heads will hook on either side of the upper nail. The whole affair will then balance on the head of a single nail.

One more? Okay. The tooth-and-nail puzzle is one you have to do at home and bring to the job. Take a 1 1/2 inch square piece of wood, about 4 1/2 inches long (basswood works best) and mark it in seven equal parts down its length. Saw and chisel out three slots that run two-thirds of the way through the block. This leaves four teeth. The challenge is to put a nail through the two middle teeth, leaving the outer teeth intact.

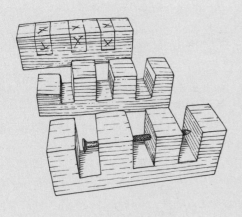

How the nail was trapped.

The trick to doing this is to get one of the outer teeth out of the way. Boil one end of the block in water for twenty minutes. Quickly set it in the jaws of a vise and compress the boiled end tooth to half its original thickness. Work slowly enough for the water to squeeze out but fast enough to get it done before it cools. After a day in the vise, the tooth will hold its new compressed shape, giving you access to drill the hole for the nail. Make the hole loose enough for the nail to move around in, and drop the nail in place. Set the compressed end back into boiling water, and in a minute or two it will expand to almost its original size. Let it dry and plane off any discoloration from boiling. The nail will look inexplicably trapped in a wooden cage.

You can try to justify these tricks by claiming that they force others to think about important things--things like hygroscopic wood expansion, centers of gravity, and earwax. But really, they're just one way we push at each other down here in the trenches: tight little packages of subversive woodworking.

7 Joiner

JOINERY, is an Art Manual, whereby several Pieces of Wood are so fitted and Join'd together . . . that they shall seem one intire Piece.
—Joseph Moxon, Mechanick Exercises, 1678

In *A Midsummer Night's Dream*, Shakespeare created the characters of the "rude mechanics," with their rudely suggestive names — among them, Bottom the weaver, Snout the tinker, and Snug the joiner. If Snug was a good joiner, his work was snug indeed.

The joiner connects pieces so that they not only fit at the moment of execution, but will continue to fit as the wood continues to move. Wood is a lively material, but joinery has evolved designs that allow it to form a stable object, such as a door or a tabletop. Still, as Shakespeare knew, there are limits to every art. In *As You Like It*, he wrote of the consequences of using wood that was still too close to the tree. "This fellow will but join you together as they join wainscote, then one of you shall prove a shrunk panel and, like green timber, warp, warp."

Snug was comically out of place in Oberon's forest, for joinery takes woodworking more than a few steps away from nature. Snug's work was likely undertaken at a shop in the village where he could keep his glue pots and wood, saws and mallets and chisels, and, most significant, his planes and benches. The material came to Snug the joiner already cut, and to some extent, dried.

Seasoning

Now saw out thy timber, for board and for pale,
to have it unshaken, and ready to sale:
Bestowe it and stick it, and lay it aright,
to find it in March, to be ready in plight.
     — Thomas Tusser, Five Hundred Points of Husbandry, *1557*

The carpenter's work may come first on a new house, but the joiner's wood was the first cut. Stock for floors, doors, windows, and trim needs this extra time to lose the water of life. With the loss of water, the wood shrinks across the grain and becomes harder, stronger, and lighter — ready for the joiner.

Seasoning is so important that it is common practice to make critical items like doors in stages, allowing for "second seasoning." The joiner brings the material almost to its final size, then sets it aside for a week or two before continuing. Oak is particularly prone to shrinkage after a fresh surface is exposed. A builder's guide of 1726 warned of this: "For it has been observ'd, that though Boards have lain in an House ever so long, and are ever so dry, yet when they are thus shot and planed, they will shrink afterwards beyond Belief." Court records show that one builder working on the College of William and Mary in 1704 was hauled into court because "the Plank & timber being green and unseasoned," the work "was shamefully spoilt."

You can't hurry wood. An experienced artisan can easily tell air-dried wood by the sweet way it cuts. Some of the best artisans I know have mastered the art of drying wood by forgetting about it. They stack it outside, protected from the rain, but open to the air. After a year or two or three, they bring it into the

Sawn and separated by stickers, the boards dry under cover for a few years.

shop and put it in the rafters. After another year or two or three, they start to look at it again. The longer they can forget about it, the better. They have no idea of its numerical moisture content, but they know when it's ready.

Still, it's the American way to hurry things along. In *My Year in a Log Cabin*, William Dean Howells recalled kiln-drying lumber for his family home in the 1850s:

> The frame had been raised, as the custom of that country still was, in a frolic of the neighbors, to whom unlimited coffee and a boiled ham had been served in requital of their civility, and now we were kiln-drying the green oak flooring-boards. To do this we had built a long skeleton hut, and had set the boards upright all around it and roofed it with them, and in the middle of it we had set a huge old cast-iron stove in which we kept a roaring fire.
>
> This fire had to be watched night and day, and it never took less than three or four boys of the neighborhood to watch it, and to turn and change the boards. The summer of Southern Ohio is surely no joke, and it must have been cruelly hot in that kiln; but I remember nothing of that; I remember only the luxury of the green corn, whose ears we spitted on long sticks and roasted in the red-hot stove; we must almost have roasted our own heads at the same time.

The heat was surely hard on the oak boards as well. Wood dries on the outside first, with moisture deeper in the wood taking longer to get out. That's what causes the checking and cracking as wood dries too fast—the inside is still fat while the outside shrinks. Slow seasoning gives the deeper wood a chance to keep up with the outside.

Wood also shrinks and swells differently in different directions. It never gets longer and shorter, just fatter and skinnier—and it does that unevenly as well. From green to oven dry, wood shrinks about 10 percent in line with the growth rings (tangentially) and about 5 percent across the rings (radially). In a log left in the round, the wood will often crack open to relieve the stress of this differential shrinkage. But if you split the log first, either with wedge or saw, the timber can open up without cracking as it dries.

A radially cut or split-out panel will get smaller as it dries, but at least it will stay rectangular. The even shrinkage of a radial, across the growth rings, "quarter sawn" cut of a tree makes for even shrinkage. Boards cut tangential to the growth rings (flat sawn) will warp as they dry, cupping on one side and crowning on the other, the rings tending to flatten out. You can plane off the cup and crown, but wood is only as stable as its environment. If the wood gets wetter or drier, it will warp again. With individual pieces getting bigger and smaller across their width but not their length, your designs must allow the wood to move.

Much of the art of joinery is devoted to accommodating wood move-ment—but seldom from dead green to bone dry. It's a matter of steps along a

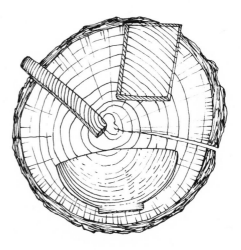

Drying wood will shrink more in line with the growth rings than across them.

path. When making drawknifed chairs, you count on the shrinkage of the posts to grip the rungs tight. Even then, it's usually very dry rungs set into less dry posts—not sopping green posts. Dead green wood responds well to splitting and shaping tools, but the tools of joinery want wood that hangs tighter.

Joiners often plane wood before it has fully seasoned—it's easier and faster—but it usually works best after the wood has made a start on the path to dryness. Bob Simms, working in an English shop in the 1920s, was always glad to work on unseasoned elm:

> We had a coffin maker next door to the cabinet shop. I often used coffin boards because you could buy what they called a "coffin set." It was rough and needed to be planed out, but they were cut to sizes for coffins. You could buy a whole set, the sides and the end pieces, far cheaper than you could buy one of the boards out of the stack in the yard. We used coffin boards because they were fresh cut and planed out beautifully. But if you dry it for a couple of years you couldn't work it with a bloody ax!

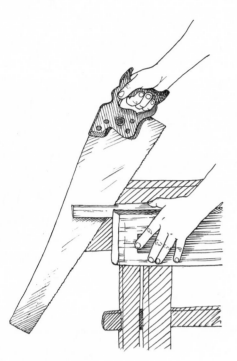

Hold the ripsaw at about 60 degrees to the surface.

Sawing Out

So now the stock is cut and dried, ready for the saw (or the bloody axe). First, however, come the square and ruler, the snapline and gauge. Spread out all the stock for the job and consider how to divide it in order to minimize both wasted wood and labor, as well as making the best matches in grain and color. Beauty and exceptional width should be reserved for a worthy setting, and unneeded segments should remain as lengthy as possible so they can be used later. Tapered and curved pieces should nest as closely as your saw will allow.

You'll find old books with advice on everything, including sawing. Professionals grew up in their trade, and few of them wrote books or learned from them, so many of the writers were talented amateurs. One such French writer from the late 1700s advised:

> Few amateurs are not embarrassed to exactly follow a line and cut straight with a saw.
>
> Fault one: Insufficient care in sharpening the saw.
>
> Fault two: Cutting green or soft wood with saw blades whose teeth are fine and lightly set—and the reverse for hard, dry wood.
>
> Fault three: The impatience of the user, who, trying to saw more quickly, presses too hard on the saw: this flexes the blade and makes it run to the right or to the left.

This gent wrote woodworking books at night under the pseudonym Bergeron. His day job was serving as the public prosecutor for the city of Paris during the French Revolution, so we'd better listen to him: "To saw well, then, choose the right saw for the wood that you are cutting and the work you want to make. Work perpendicularly and parallel to the line, and, from time to time, lubricate the blade with bacon or tallow."

The right saw depends on the direction of the grain. Across the grain, you use a crosscut panel saw sharpened with knifelike teeth. Down the grain, you use a ripsaw with chisel-like teeth. Cutting diagonally, "a rip saw cuts faster, but a crosscut, smoother."

From a distance, hand ripsaws look the same as crosscuts, but rips run larger in the blade and coarser in the teeth. Sometimes you'll see a number stamped on the inside heel of the saw indicating its coarseness in points to the inch. Points are literally points and not entire teeth, so there is always one less tooth than points in the measure. Ripsaws range from four to seven points and crosscuts from six to twelve.

More teeth make slower but cleaner going, and some saws are cut with more teeth at the toe end of the blade to make starting the cut easier. Assuming you are right-handed, hold your left hand around the end of the wood and place your left thumb against the toe of the blade of the saw to hold it steady. Draw the blade lightly back and forth and the cut will begin.

Once the cut is started, work with a crosscut saw inclined at 45 degrees to the surface and a ripsaw at 60 degrees. Depending on what you're out to accomplish, you may want to saw right on the line, saw down the side of the line, or split the line with the edge of the kerf. Since we're roughing in the stock, this is beside-the-line-with-a-little-clearance-on-the-waste-side sawing.

Should the saw veer away from your line, lower the angle of the saw to the work surface and bend it slightly along its length to move the kerf in the right direction. When ripping or crosscutting thick stock, first saw diagonally into the corner and then bring the saw around square to finish the cut. When ripping, turn the piece over from time to time and saw from the other side. In critical work, turn the piece more frequently. As a cabinetmaker friend put it, "Saw only the lines that you can see."

Because any long curved piece will be aligned down the grain for strength, saws for curves generally have rip teeth. Curved kerfs also call for narrow blades. A compass saw is a narrow and stiff handsaw, good for cutting gentle curves like tabletops. Tighter turns need an even narrower blade. Turning saws can cut tight curves in relatively thick planks of wood. As with a coping saw, the knobs can turn the blade at angles to the frame, allowing it to pass beside the work as you cut.

While the bow saw uses the twisted cord to tighten the blade, a frame saw uses the blade itself to hold the works together. By stretching the blade between the ends of a wooden rectangle, you can tighten it with wedges or a wing nut arrangement. Since the blade is in the middle rather than on the side of the rectangle, this saw works well for ripping timbers, the frame passing on either side of the wood.

The most common frame saw is the wheelwright's felly saw, used to cut the arcing segments of the rim of the wheel. Wheelwrights generally work with the wood held vertically, as do veneer sawyers, but old images of joiners ripping regular stock show the wood horizontal, the vertically held saw benefiting from gravity on the down stroke.

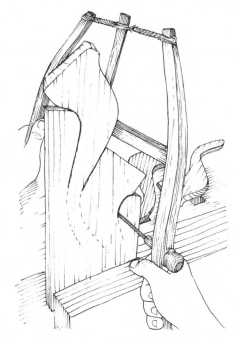

The turning saw can cut the tight curves for the legs of a pedestal candlestand.

The frame saw does better in heavier stock.

The Nib and the Blade

Good handsaws are taper ground: that is, the metal is thinner on the back than it is on the tooth side to permit an easier passage through the kerf. They may also be ground thinner near the toe end of the back than at the handle end. You can test the grinding and temper of the steel by bending the blade in your hands. The evenness of the bend, the stiffness of the return to the flat, and the tone of the blade when struck with your thumb reveal the quality of a saw.

It can also reveal if your fingers are too slippery, as Bob Simms found out when testing a saw in 1922.

Did I tell you the story about the tool shop? The rule is, when you go to buy a saw you don't just pick a saw, you bend the tip around and pass it through the handle and if it doesn't come back straight — you throw it out. So I went in and said I wanted two of these handsaws and while I was testing one, it slipped and went right across his face! These were times of depression, you know, and he was so anxious to do business that he never even murmured! I grabbed it and the one next to it and got out of there! I still dread that saw.

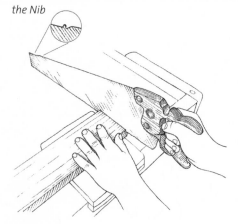

the Nib

Ponder the nib as you hold the wood in the bench hook and saw across the grain.

The nib on a saw poses an even more vexing question for many. You won't find them on today's handsaws, but until the mid-twentieth century, saws commonly had a tiny protrusion, a "nib," near the toe end. So, what's it for? There are two schools of belief. Existential Nibonists say it's a vestige of the decorative work found on early Dutch handsaws and serves no purpose. Utilitarian Nibonists say the nib must have a purpose, and are divided into warring subsects. Guardians say the nib is there to secure a wooden blade guard. Startarians say that it's an aid to starting the cut. Temperance Utilitarian Nibonists believe that it was used in the heat-treating process.

I've seen it happen too many times. Galoots gather, liquor flows, and the tooligans start picking fights. I used to be a club-swinging Existentialist myself, but after taking a few skull-cracks, it became clear to me that the nib was indeed put there for a purpose. Nibs were put on saws — just to mess with our heads! To drive us apart — when hand tool users should be united!

Sharpening Saws

You might skip over this part until you calm down, unless you find, like me, that the concentration required for saw sharpening can help you regain your peace of mind. There are four steps in sharpening a handsaw: jointing, shaping, setting, and sharpening.

Jointing is the process of making sure that all the teeth are at the same height. Clamp the saw, teeth up, between boards in a vise or in a special saw-sharpening clamp. Take a fine flat file and lightly drag it square along the length of the teeth precisely perpendicular to the side of the blade. Joint lightly until you brighten the tips of all the teeth. One or two passes should be enough.

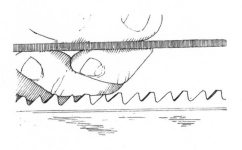

Jointing levels all the teeth of the saw.

Shaping, the next step, puts the teeth at their appropriate profile for their allotted task. Both crosscut and rip teeth are shaped with the same equilaterally triangular file called a "slim taper." The 60-degree angles of the corners of this file can cut both shapes of teeth, the difference being made by the axial tilt given to the file as you work.

Finer saw teeth call for finer files. Saws with four to eight points to the inch need a six-inch slim taper file. Those with nine or ten points need an extra slim, and a double extra slim file will be required for saws with 11 or 12 points. Even more finely toothed saws will need a five-inch superfine file.

Rip teeth are filed so that their cutting face is at 90 degrees to the line of the tops of the teeth. Thus, if the handle of the saw is to your left as you are filing, you shape the teeth with the left face of the triangular file held vertically. Crosscutting teeth, however, are shaped with a 12- to 15-degree slope off the vertical on their faces, and the file must be held accordingly.

Shape the teeth on a ripsaw by filing straight across without dipping the file on either side. On crosscuts, you can hold the file at the sharpening angle, but concentrate on the shape of the teeth, not their acuity. Start at one end on the face of a tooth that leans away from you and file it and the back of the adjacent tooth until you reach the middle of the flats caused by jointing. Skip to the next gullet where the face of the tooth leans away from you and do the same thing. Work your way down the length of the saw, doing every other gullet and then reverse the saw in the vise and repeat the process down the other side. On this second run, the filing should take off the remaining half of the jointing flat, and the teeth will all be sharply pointed and equal in depth, height, and angle.

Once the teeth are all the proper size and shape, they can be accurately set. Setting the teeth means bending them slightly to alternate sides, right and left in turn. The set gives the blade clearance in the kerf and makes a huge difference in how the saw performs. The maximum amount of set to put into any given tooth is to have it lean one-third the thickness of the blade to the side, bent at a point halfway down from the tip. You may want this maximum amount for coarse cutting in green wood and considerably less (or none) for dry. You can always add more set if you don't have enough.

There are numerous devices for setting saws, some easier to use than others. Patent pistol-lever saw sets are the easiest to use, as well as the easiest to overset the saw with. If you don't have the instructions that came with such a set, experiment on your saw at the handle end, starting with what appears to be the minimum adjustment. Set the teeth from both sides or you won't be able to tell what the total effect will be. The set must be equal on both sides of the saw or it will tend to pull to the side.

The saw wrest is a wrenchlike slotted bar used to bend individual teeth or to twist adjacent pairs of teeth in alternate directions. By setting the slot on the tops of two teeth and turning the handle to the side, you bend one in one direction, the other in the other direction. You work your way down the length of the saw, two teeth at a time. The same saw wrest will also let you bend just individual teeth right and left.

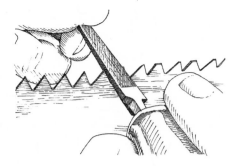

Shaping corrects the profiles of the teeth.

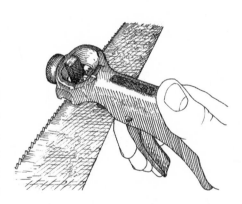

Setting bends the teeth to alternate sides, keeping the saw from binding.

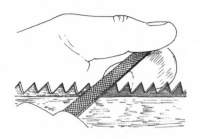

Sharpening puts the edge on the teeth. For crosscutting teeth, work on the face of a tooth that leans away from you, except now angle the handle of the file back at 45 degrees toward the handle of the saw. Don't dip too much, just angle back. File the face of one tooth and the back of the adjacent tooth simultaneously until you form half of the point on each of them. Do every other pair down one side of the saw, and then flip the saw around to do the remaining teeth.

After setting, rip teeth may no longer be properly oriented. Bending the teeth to the sides cants the angles of the tops, and twisting the teeth cants the angles of the faces. You may need one more light jointing pass and then a light shaping pass from opposite sides of the saw.

As a final touch on crosscuts, lightly pass a whetstone held flat down both sides of the blade to ease the burrs from filing. I have always sharpened by filing the faces of teeth that lean away from me, going with the set of the teeth. Filing against the set leaves less of a burr, but the final stoning always takes care of that anyway.

Sharpening gives each crosscut tooth a knifelike edge.

Bench, Stop, and Holdfast

You can saw on horses, but the plane needs a bench. The plane, the bench, and the stop probably all arrived at about the same time—for one does little good without the others. Unless supported by a bench top, light wood will bend away from the plane, and unless there's a stop at the end of the bench top, the wood will fly off the end with the first stroke.

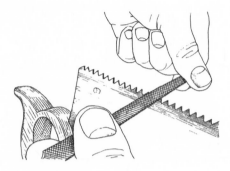

Sharpen ripsaw teeth square across.

André Roubo, whose Parisian wooden dome inspired Thomas Jefferson to poetic reverie, illustrated a joiner's bench that is now known by the great French joiner's name. A Roubo bench has legs attached to the massive top with characteristic dovetail and tenon joints. There is an adjustable stop through the top and a fixed stop on the left front leg. Directly beneath the top is a drawer and a little swing-out tallow pot. A shelf for tools rests between the stretchers that strengthen the legs.

The top and face of the Roubo bench are also pierced by holdfast holes. The L-shaped holdfast is brilliantly simple. Set its shaft in a slightly oversized hole, bonk its top with a mallet, and it becomes "cocked" in the hole, holding the pad down with the force of the captured hammer blow. To release it, just tap it on the back of the head and it springs free.

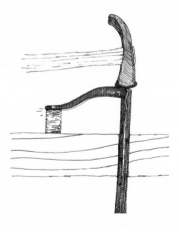

A tap of your hammer "cocks" the holdfast in the slightly oversized hole.

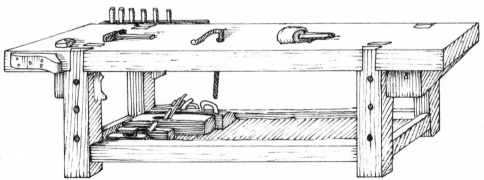

The bench illustrated by André Roubo has no vises.

Roubo bench tops are single massive timbers, two feet wide and as long as needed. It's the sort of plank you might split and hew from a massive storm-felled oak, but timber this wide is certainly harder to get. I have seen a late nineteenth-century French workbench made by using narrower heavy timbers, the width filled out with a tool well or till. The back legs are still tenoned through the narrow top timber but maintain a wide stance by splaying out to the back.

These later benches also have screw vises instead of stops on the left front. Roubo shows a screw vise on the leg, and Francis Nicholson's 1812 illustration shows a bench with a screw vise passing through the front skirt.

We see Nicholson-style benches in many British genre paintings of the Romantic era, but the bench that became the dominant form is neither French nor British. The key component of the German bench, as illustrated on the opening page of this chapter, is the tail vise, a screw-operated box moving in a track. Fitted with moveable dogs, the tail vise can pinch a piece of wood on the bench top and hold it for planing. None of the surface has to suffer the damage or intrusion of the holdfast. The end vise also gives another place to work, clamping wood vertically between its jaws.

Brilliant work has been done on all of these benches, but the advantages of the German bench were immediately recognized. Yet with all their accessories that make them handy for sawing, chiseling, and such, their essential function for the joiner is giving a level and solid backing for the work of the plane.

As Bernard Jones wrote in *The Complete Woodworker*, "Planing up is the actual foundation of a job." Small, unintended twists and tapers become magnified as you build. If the work is to "seem that it is one intire piece," only properly dimensioned stock can grow into tight joints and flat frames.

Planing is not just a good start but also the best finish. A finely set, well-sharpened smooth plane takes a gossamer shaving and leaves behind a gleaming

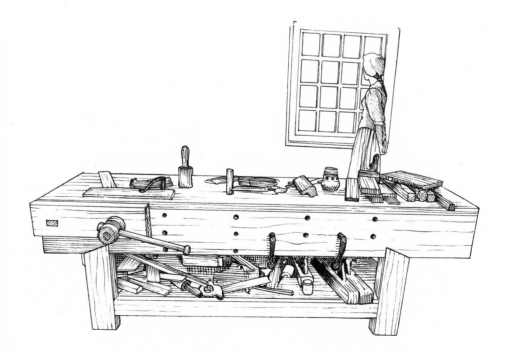

An English-style bench as illustrated by Francis Nicholson in 1812.

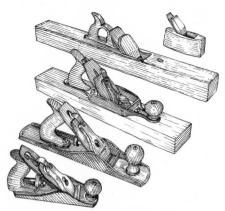

Front to back: small iron plane sharpened
as a scrub; iron jack plane; transitional trying
plane; wooden jointer; wooden smooth.

surface. But trying to do all your work with that finely set smooth plane would take forever. That's why the joiner employs a trio or even a quintet of bench planes. Just as the speeds on a bicycle match the rider's strength to varying terrains, the joiner's planes make progressively finer and broader cuts as the work progresses. Trying to bring rough stock to precise dimensions with a single plane is like riding a one-speed bike in a hilly town.

I'll stretch the hill analogy to serve for the function of planes as well. A plane surface has no hills, no valleys or twists. Push a long plane down an uneven timber and the plane's body will span over every valley, allowing the iron to cut off only the hilltops. Subsequent passes with the plane continue lowering the hilltops until they are no more.

The plane iron (as the cutting blade is called) not only levels the hills, it also leaves a shallow valley with each passage. Unlike chisels, bench planes usually work with many overlapping strokes on surfaces that are broader than their blade. If the corners of the plane iron were square, like a chisel, each stoke would leave a square valley with ragged margins. That's why a plane iron's corners are rounded to varying degrees—to make the depth of cut diminish to nothing at the margins. You start leveling a surface with a plane bearing a deeply protruding, heavily rounded iron, and each plane in the progression takes a finer cut with a more squared-across iron.

Sharpened and Set

Plane irons may benefit from rounding across their breadth, but rounding on the bevel edge or flat face means they are dull. Bench plane irons work with their bevels down toward the wood. To see the problem with a rounded bevel, put your index finger on the table and push it forward as though it were a plane iron. Unless your fingernail is much longer than mine, its edge will not make contact. This is exactly what happens with an ill-ground or dull plane iron—it rubs and nubs but never cuts.

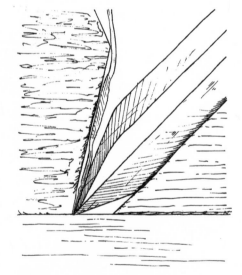

The cap iron works with the narrow mouth
of the plane to ensure that the shaving is
only cut by the edge—never split by the
wedge.

Before you can sharpen the iron, you have to take it out of the plane. Iron-bodied planes usually have a cam-lever that locks the iron in place. Pull the lever and the iron comes free. On a wooden plane, strike the fore end of the top of the body sharply with a wooden mallet as you support the plane with your free hand. Often a wooden plane will have an inset "start" or "strike button" at this point to save the body of the plane from wear. On small planes the strike button may be on the rear, and that is where you should tap to release the iron.

I say "iron," but you probably found two irons, back to back, held together by a screw. The lower one is the plane iron, the upper one the cap iron or chip breaker. This upper iron is an innovation from the late eighteenth century to help the plane leave a smoother surface. When a shaving is severed by the edge, it slides up the face of the iron and can build enough leverage to lift a shaving ahead of the iron. This lifting can cause the wood to split a tiny bit deeper into the surface than the iron can cut. The chip breaker prevents this by sharply turning the shaving and breaking it before it can do any harm.

The cutting edge of the plane iron and the breaking wedge of the cap iron work as a team, but we need to separate them for the moment. Loosen the screw holding them together and remove the cap iron, turning it sideways and carefully avoiding touching and harming the cutting edge. This care to protect the edge is even more critical after sharpening. One slip and you're back to the stone.

The shape and bevel of the plane iron come from the grindstone. For a plane iron you plan to hone with a secondary bevel, grind to 25 degrees. Grind at 30 degrees if you are going to hone the whole bevel. You can always tell a 30-degree angle by the length of the bevel. When the bevel is twice as long as the thickness of the iron at the end of the bevel, you've got 30 degrees.

The flat side of the iron is where the steel is. In older plane irons you can see the color change or the weld where the steel layer ends. The flat side of the iron must remain flat right up to the very edge. It should never need grinding—just a dead flat polish on the whetstone.

Check the iron with a square before and as you grind to help you judge the degree and evenness of the widthwise curvature. It's a very slight arc on a jack plane, just 1/16 inch at most. Smooth, trying, and jointer planes just get the very corners rounded off. Just bear down a little more on the ends of the iron as you sweep across the grindstone, and, later, across the whetstone.

The cap iron needs no attention unless you find shavings catching under it. It must bed tight against the flat of the plane iron with a smooth face ready to give a quick turn to the shaving.

Carefully reassemble the irons with the big screw. A jack plane may have the breaker as far back as 1/16 inch; a smooth or jointer should have a setback of down to 1/64 inch. The smoother and finer the cut, the closer the breaker iron should be to the edge. Difficult wood needs the breaker set as close as possible to the edge. In nicer stuff, you can back off the breaker to make the cut easier. I have a very old and beautiful wooden single-ironed jointer plane that is a joy to use. It seems to glide down the edges of boards because there's no chip-breaking work adding resistance.

Before the iron goes back in an iron-bodied plane, look at the sloping ramp where the iron rests. This is called the frog, and if there are two screws holding it to the body of the plane, it is an adjustable frog. If you loosen these screws, you can use the screw behind the frog to adjust the mouth, the opening in front of the iron. For fine shavings, narrow the mouth. For coarse work, widen the mouth so the fat shavings can pass through. You'll need to tighten the frog screws and lock the iron in place before you can check the mouth opening. Work by trial and error, taking great care not to touch the freshly sharpened iron against the metal body of the plane. All-wooden planes don't come with adjustable frogs, but they as well as iron-bodied ones benefit from narrow mouths. That's how planes eliminate the wedge effect, ensuring that the bottom of the plane holds down the wood until the last possible instant before the edge of the iron shears it off.

When you try the plane, you may hear and see chatter. Chattering occurs when the iron repeatedly bends back and then springs forward as the plane progresses along the wood. The iron is either too far extended or is not well

Loosen the two screws in the frog so you can adjust the mouth opening ahead of the iron.

supported. Very thin irons or long sharpening bevels are also prone to chatter. Either reduce the amount of protrusion of the iron or correct its bedding on the frog.

Setting a metal plane is as easy as turning a knob. Hold the plane upside down and sight along the bottom. Thumb the depth-adjusting nut around until the iron shows its head. Push the lateral adjustment lever left and right to even the iron. Now turn the depth nut back to take the iron out of sight. Turn it back again until enough iron shows to take a paper-thin shaving. If you go too far, back up and come out again. To take up play in the mechanism and make the iron stay where you want it, always make the final adjustment in the direction that moves the iron outward.

For a wooden plane, it's just tap, tap with the mallet or hammer. Place the iron back in the plane, bevel down, and push the wedge into its seat. Turn the plane upside down and sight along the sole to judge the squareness and depth of the iron. Set the iron a tiny fraction of an inch shallower than you want and lightly tap the wedge home. The iron will usually travel a bit deeper with the wedge. If the iron appears to be too deep, a sharp tap on the heel end of the body will raise it. This loosens the wedge as well, so it will want another tap. If the iron is too shallow, tap either on the front end grain of the body or on the top end of the iron itself. The wedge may again be loosened by tapping on the iron, so see that it is well secured before you test the plane. You can make all the adjustments on a wooden plane with a mallet, but a hammer on the top of the iron works faster and finer.

Tap the nose of a wooden plane to make the iron take a deeper cut.

In the long run, finer and faster go together in planing. As Bergeron advised in 1796: "Often an amateur tries to advance his work by giving too much iron, but in little time his plane is choked. He is obliged to withdraw the shavings with an iron point; thereby he destroys the plane's mouth, nicks the iron, and loses much time."

Scrub Planes and Winding Sticks

I shall cleanse on every side
To help my master in his pride.
    —Debate of the Carpenter's Tools, *1500*

The art of the plane would seem to be all edge—no wedge. The wood is cut so finely that it never builds up enough leverage to split, even minutely, ahead of the edge. At least that's the theory. In practice, we use a sequence of planes moving closer to that ideal.

Working across the grain with a short scrub plane, the joiner can quickly level the surface of the wood. The jack or fore plane (15–17 inches long), working down the grain, smoothes the rough hollows. The trying plane or the jointer plane (18–30 inches long) renders the surface dead level. When the shavings emerge unbroken for the entire length of the stock being planed, the surface is level and true (or close to it). The final plane, the little 8-inch long, smooth plane, is set extremely fine and gives the final touch.

Use a plane with an iron rounded across its width to scrub across the grain.

If your stock is rough sawn but otherwise civilized, you could start planing with a jack. But when you have some wide, wild timber lying on the bench, you want to start with a scrub plane. The scrub is a short plane with a big mouth and a rounded iron used in short strokes across the grain—like you're scrubbing a floor.

On a very heavy piece with significant distortion, say, some big hewn timber for a workbench top, you'll have to establish level lines to guide the scrubbing. Set long, straight sticks across opposing ends to help you scope out the twist. You'll have to take this wood off anyway, so cut a narrow shoulder on each end of the timber for the winding sticks to rest on. Plane or saw and chisel away the high corners of these shoulders, testing and making adjustments until the winding sticks sit in alignment. Connect these aligned shoulders with chalk lines snapped down the sides and you have defined the line between wanted and unwanted wood.

These guidelines will be easier to follow if they are physically cut into all the shoulders of the plank. Rabbet planes are made just for such jobs. You might also want to carefully make a series of saw cuts across the face stopping just shy of the chalk lines on the sides. If there's a big hump to remove, make enough saw cuts through it so that you can safely take it off with a mallet and chisel without creating deep tear-outs. Be careful—a heavy hand with the chisel can easily make much more work for the plane.

The scrub plane hogs the wood off fast—sometimes too fast. It cuts narrow furrows across the work, often leaving it rough and torn. Take care that you don't tear out a big chunk when the plane reaches the far edge, perhaps ruining the corners. You also have to stop before you push this rough surface too close to the bone and leave tear-outs that require excess finish planing to remove.

Holding a plane askew to the direction of the push makes it take a narrower shaving. This makes the plane easier to push but also shifts the carefully worked out geometry of the tool. It lowers the attack angle of the iron, which also makes an easier cut, just as it's easier to walk diagonally up a slope rather than straight up. The skewed blade works with a sliding cut, like slicing bread by pulling back on the knife rather than just pushing straight down, and that's easier too. A skewed push also widens the mouth ahead of the cutting edge and shortens the bed of the plane. With some combinations of grain direction, moisture content, and plane geometry, you may find skewed planing to your advantage.

Jack in the Country

The jack plane can work across the grain to start, and then gradually turn with the grain as the work begins in earnest. Here's where you learn the truth of the grain. If the surface is rough and jagged, you're working against the grain. Orient the wood on the bench so you're planing into the backside of rising grain, like stroking a cat from nose to tail. In some pieces you can have rising grain on the left half and falling grain on the right half. The grain will also rise and dive around a knot, so you may find places you can only plane into or out

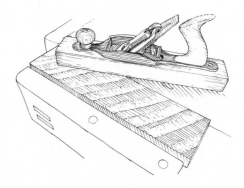

Follow the scrub with a jack and a trying plane to level the hollows.

of, but not straight across. Here again, tear-outs can make extra work for you. Turn either the wood or the plane, and back off the iron to take finer shavings. Wild wood can be a challenge, but clear, air-dried, straight-grained stock will work sweetly with shavings flowing from the plane in long rolls.

The jack plane has a handle or "toat" for your right hand, but no specific place for your left. When you're planing a broad surface, you lead with the inside of your left wrist. Place your thumb on the near side of the plane and grasp the far side with your fingers. Many old planes have deep contours worn into the beech by the long-dead user's fingers.

When you're planing the edges of boards—when the wood is narrower than the plane iron, you can turn your left wrist around and grasp the near side of the plane. Place your thumb on top and let your fingers ride against the side of the board to steady the plane in its passage.

The jack works with the grain in shorter strokes than the longer planes that will follow. Work your way to the end, taking off the obvious high spots as you go. You'll find wax or tallow rubbed on the bottom of the plane a great help. As with any plane, ease off and give the iron a little lift across the surface on each return stroke. Keep sighting down the surface to see how it's going. Check for wind with the sticks, and for cupping or crowning with a straightedge. When the jack plane has reached the entire surface, you may want to adjust it to take a finer cut and have another go, or it may be time move on to the next plane.

When you pick up the next plane, you must put the first one down. I loyally attend the always-set-the-plane-down-on-its-side school of thought. This may not be as important in pristine workshops, but if you work near any foot traffic on bare earth, the air is constantly full of settling grit—making every horizontal surface death to the cutting edge.

Trying Plane and Long Jointer

The master tosses a long plane to one of the apprentices and shouts, "What is this?"

"A trying plane!" she responds.

"No!" The master snatches it away and tosses it to another apprentice, nailing him in the solar plexus. "What is this?"

"A jointer!" he gasps.

"No!" bellows the master, "Trying plane and jointer are just noises! This is not a noise! Here is what this is!" The master grabs the tool, turns to his bench and makes a single long shaving down the edge of a ten-foot cypress board.

There's an overlap between the names for planes in the 20- to 30-inch range, but the master's last act makes the plane a jointer. Jointing refers to making the narrow edge of a board perfectly straight in preparation for edge-to-edge joints. You don't think of "jointing" a broad surface, but "trying" fits the job just fine. To me, then, it's a jointer when jointing the edge, and a trying plane when leveling the corrugations left by the jack plane.

A long plane by any name will level a surface. The long sole spans any low spots, holding the iron above them until the last of the risings are gone. These

Winding sticks will amplify any subtle twist in the surface.

planes must traverse the whole length of the surface with each stroke, so instead of standing in one spot as with the scrub and jack, you walk the plane from one end of the piece to the other.

You can hear the progress of the plane as you work. On each pass down the board the silences as the iron rides over hollows grow shorter. The shavings become longer and longer until the plane makes one constant cut down the entire length of the board. When you get the one unbroken shaving, the surface is either dead straight or in a long, gentle arc.

The arc might be caused by the plane. Tapping the wedge a little too hard into a wooden plane can spring the body into a gentle curve, but that is rare. It's usually the planer and not the plane.

The arc arises, as Bergeron wrote,

> because the wood is easier to cut at the beginning, and the right hand grasping the handle of the jointer pushes down on the part of the plane which overhangs the end of the board. In the same way, when one pushes the jointer to the other end, it is the left hand pressing on the fore end, which imperceptibly dips the tool. These effects, though inconsequential in themselves, become serious when they are multiplied by the number of times that the jointer passes over the piece.

To keep this from happening to you, press down only with your left hand on the fore end of the plane at the start, and only push forward with the right hand. The stroke starts with just the part of the sole ahead of the iron in contact with the work. Slide forward, bringing the iron into the cut and follow through all the way off the far end. Push down with even pressure through the middle. When you get to the far end, stop pushing down with your left hand, and finish with pressure from the right hand alone. The iron is clear of the wood before and after each stroke.

Smooth Plane

The short scrub plane started us off, and the short smooth plane will help us finish. The little smooth plane is sharpened and set the same as the long finishing planes, but can come at contrary grain from whatever angle seems to work. Adding to the virtues of their small mouths and fine cut, some smooth planes (and others) may have their irons bedded at 50 degrees, rather than the common 45. This is York pitch, and the steeper angle helps in gnarly hardwood.

Whether you use a coffin-shaped wooden smoother, a rosewood-infilled Norris type, or a polished-up Stanley, take the time to fully sharpen and adjust it. Don't stop tuning it, testing it, sharpening it, until you're planing bird's-eye maple with shavings so fine and thin they have only one side. It's a rush.

The finely set smooth plane gives the final touch to a surface leveled by the longer planes.

Try Square and Gauge

All this planing has given us one flat face. But joinery is built on rectangular stock, so we have three more faces to go, five more if we count the ends. There is

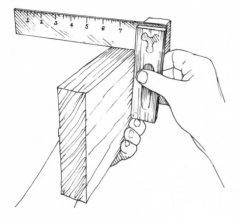

Bring the face edge perpendicular to the face side, checking with the try square.

some debate on the proper sequence for squaring up stock. Rather than indulging the narcissism of small differences, let's just get to work. Dimensioning stock with six surfaces (including the ends) follows this sequence, with the last two steps interchangeable:

Face Side
Face Edge
End Grain   End Grain
Back Edge
Back Face

Choose the face sides by laying out all the pieces and playing with them to find the most harmonious combination for the eye. You can change your mind later as planing reveals more of the grain and color masked by the rough-sawn surface. Set the face side up on the bench and pinch it with the dog of the tail vise, or set it against the planing stop and a couple of pegs let into the holdfast holes in the bench top. Shim it if it won't sit level.

The work begins on the face side. Bring the face side true with your planes and choose the face edge—the best edge to show to the world. Mount the piece with the face edge upward on the top or on the front of the bench. The edge is usually too narrow for any cross-grained scrubbing, so start with the jack and work finer and finer. Test the face with the winding sticks, the straightedge, and the try square.

The try square itself may need testing from time to time. Even the most beautiful rosewood-stocked, brass-eyed try square can become false if allowed to get wet. Test for honesty by using the square to draw a line at right angles across a straight-edged plank. Flip the square and see if the line is the same. If not, confirm that the "straight-edged plank" is exactly that. If it is, the square is out by half the divergence of the lines. Bad squares seem to just hang around. No one can redeem a false one—but no one ever throws one out.

To this bad habit, we must add another—sliding the try square down the plank. (I think this is how Harrison Ford did it in *Witness*.) The approved way is to set the stock of the try square firmly against the face side of the work and move it down to bring the blade against the face edge. Sight beneath the blade for light. Again, don't slide the try square like a scraper; instead, make repeated tests at discrete points down the work. Test and plane until you have the face edge true and square to the face side. It's customary to now mark the face side with a looped pencil mark connecting with the point of an inverted V on the face edge.

End grain planing is often unneeded if the ends of a piece are going to become a tenon or be otherwise occupied. If the end grain does need to be planed true, now is the time. Planing the end grain is tucked into the middle of the dimensioning process because the long grain planing of the back edge and back side can remove any roughness on the trailing edge of the plane stroke. I'll save the end grain for the next section and carry on with gauging and trying the four long sides.

With face and edge established, gauges give us the thickness and width of the stock. Look at old gauges with their stems deeply worn on their leading edge from rubbing against the wood, trailing the marking spur behind. You can tell which ones were habitually used right-handed and which were used left-handed. But this assumes that the owner always used his gauge by pushing it down the wood—there may have been as many pullers as there were lefties.

No matter which way you hold it, the gauge is always consistent in its task. It will always mark a line parallel to an established surface. Set it with your rule to the desired width (or thickness) of the piece and run the four lines with the gauge. Plane and test this surface and then do the fourth one the same way. All the wood has to come off anyway, but working in a side, edge, edge, side sequence lets you reduce the width of the final face planing to the minimum.

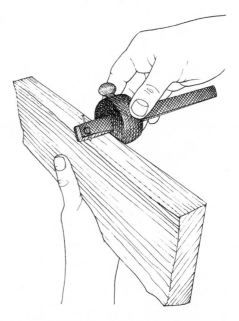

Gauge from the face side to define the parallel back side.

Block Plane

End grain is challenging to cut with an iron bedded at 45 degrees. On end grain, there's no chance of grain splitting out ahead of the edge. But even when the edge is sharp and finely set, a common plane can still dig in, pull out fibers, and chatter across the wood. You need an iron with a lower attack angle, more perpendicular to the end grain surface. You need a block plane.

Perhaps block planes got their name from working the end grain of butcher blocks, or perhaps from "blocking in" the ends of boards, cutting them to fit in a given space. In any case, the old wooden strike block planes lowered the pitch of the iron but consequently had to use a longer, thinner, and more fragile bevel to maintain clearance. When iron-bodied planes came along, however, someone realized they could make a plane with an iron bedded all the way down to 15 degrees or less by turning the iron bevel up.

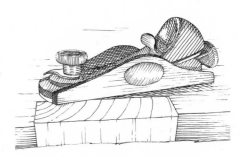

The low angle of the block plane keeps it from chattering across end grain.

It's odd, though, with a 30-degree upward-facing bevel and 15-degree bed, you're back up at the 45-degree angle of the common plane. Yet even a common block plane works better on end grain. Of course, it's the support of the edge and not the angle that makes a block plane different. Still, it always appears to be one of those phenomena jokingly despised by the French—it works brilliantly in practice, but fails miserably in theory.

Iron block planes are about the cheapest and most abundant planes out there. Many have a sliding front sole that allows you to close the mouth almost entirely. With a sharp iron, the block plane can shear off end grain in a fine, unbroken shaving, leaving a polished surface—until it reaches the end of the cut. Those last fibers on the far edge of the board have no support to keep them from splitting away, and they often do.

Putting end grain planing into the middle of the dimensioning sequence is one tactic to deal with this—you let the end splinter away, but only in wood that you're going to remove later. Beveling off the far end will prevent splintering, as will planing from both ends toward the middle. A sacrificial support at the far end will do the job as well. Just position a separate piece of wood behind the work piece to support its end grain. You can clamp this backing piece in place, or hold it by hand, both pieces pressed against the back stop of the bench hook.

The bench hook is more associated with sawing on the bench top, but if squarely made, it works as a shooting board for end grain as well. The bench hook is simply a board with battens fastened across opposite sides of opposite ends. One batten hooks over the edge of the bench, and you push the work against the other batten. One hand holds the work and the other holds the saw or the plane.

The top batten of the bench hook usually stops short of the edge so your saw can continue through the work and cut into the bench hook to leave the bench top undamaged. For end grain planing, you can make a bench hook with the batten on the flip side that continues to the end, or use a separate backing piece held between the work and the back of the hook.

In either case, hold the work piece in the hook with the end barely hanging over the edge. Set the block plane on its side, directly on the bench top. This holds the plane square to the wood as you slide it back and forth, the far end supported by the back batten of the bench hook or by the sacrificial piece.

In all of this planing with wood or metal tools, don't deprive yourself of the benefit of tallow, beeswax, or a rub of the candle to grease the skids of commerce. You don't want to leave any grease or wax on the wood that might interfere with subsequent finishing, but without something to cut the friction, the wood may be too sweat stained to take a finish anyway.

Edge-to-Edge Joints

The excellence of the flat, straight surface is that it will fit perfectly tight against any other flat, straight surface. Flooring is a simple example—one that a carpenter or joiner might undertake. Each board fits tight against the other, and the floor lies flat and smooth—but only if the joists are level and the boards are all the same thickness. If the joists are uneven, the floor will undulate. If the floorboards vary in thickness, the floor will be jagged.

It's relatively easy to level the tops of floor joists with chalk line and adze, but consider the labor required to hand plane all the boards for a floor to the same thickness. The trick is to adjust the thickness of the floorboards only where they need to be even—where they cross a joist. That's why the underside of a floor from the hand-plane era looks so uneven. The boards are all the same thickness where they cross the joist, but between joists—who cares?

Here are the steps on the floor. The boards come from the sawyers and find a place inside to dry so they won't shrink and pull apart after they are laid. When the boards are judged to be seasoned, the carpenter or joiner begins planing. Each board gets flipped over and over and the face side chosen and planed, then the two edges jointed, and nothing more. The faced and edged boards go into a pile. The boards were sawn at different times, perhaps by different crews, and their thickness might vary as much as a half inch. The joiner determines the lowest common thickness of all the flooring and sets a marking gauge

A flex-bottom compass plane can smooth both convex and concave surfaces.

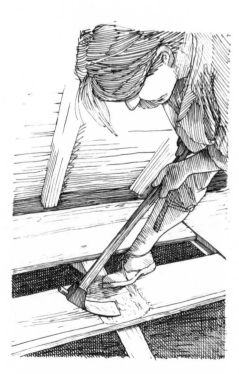

Adze floorboards to their lowest common thickness only where they cross a joist.

accordingly. Gauging down from the face side, each board gets this thickness scribed down both edges. The board now goes on the bench with the rough side up. With multiple passes of the rabbet plane along the edges, the joiner makes a narrow shoulder, or rabbet, reaching down to the scribed line.

When all the boards are gauged and rabbeted, it's time to lay the floor. As each board gets set into place, rough side up, the carpenter or joiner can see where a joist passes underneath. In those spots alone, the adze makes short work of bringing the board down to the thickness indicated by the gauge lines. Flipped over, the fat on the floorboard hangs down and the tops are level.

When four or five boards are ready for nailing, they can get pushed together as tightly as strength permits—but that's still not tight enough. Only the outermost board of each set gets nailed down, and that only after it has been moved in about a half inch. Now, the inner boards don't have enough room to fit back in unless two of them are folded upward, tentlike, against one another.

Hearing the call, the other workers walk over and give a big jump on the fold, forcing it flat, squeezing the boards tightly together. Of course, the other guys left their hammers back where they were working earlier, so they can do nothing but stand and talk and smoke while the guy laying the floor crawls about, trying to get it all nailed down before they wander away.

Nail boundary boards a bit too close together. This lets you spring the boards between them tightly into place.

Rubbed Glue

If you look at the end of a board and see that the rings run across the width, that board will warp. Perhaps not much, but in humid weather the rings will curl tighter, and in dry weather the rings will flatten out. Since the warping is predictable, you can arrange the boards to cancel out the cumulative effect. Alternating the heart side down and heart side up will ensure that one cups up while the next cups down in a gentle undulation.

Clamping two boards together and jointing them at the same time is another error-canceling practice. Set the boards together, face to face, with the edges you're joining facing upward. If you now shoot these edges simultaneously with the jointer, any tilt of the plane will place identical slopes on both pieces. Folding this joint together will make the edge slopes cancel each other out and the broad faces lie flat. This practice has a price. Planing two boards at once cancels out angular error but doubles crown error.

The rubbed glue joint permits neither error. If you've ever felt how tightly two wet panes of glass will stick together (they have to be slid apart), then you understand the power of the rubbed joint. Like the glass, the boards in a rubbed joint must be jointed dead flat. You can test the accuracy of the jointing by clamping one board edge up on the face of the bench and setting the other atop it. You can see right off if the faces are not in the same plane or if there are gaps along the way. Pivot the top board around its middle to see that it drags evenly. If the top board pivots outward with too little friction, there is crown in the middle. Too much friction or catching on the return to alignment means the middle is hollow.

This testing works only with boards that are too stiff to be affected by gravity.

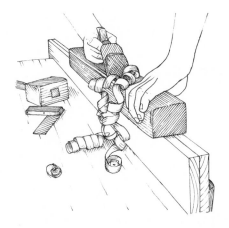

The excess thickness of gauged and adzed floorboards hangs to the underside.

Plane the joining edges of boards while they are clamped together. Any angular error will cancel out when the edges are folded together.

In building up a table top of very long, narrow boards, each pair will gain stiffness as they are joined. Pairs are then edge-planed together and tested before they get joined into quartets and so forth. Long boards and large work will need clamping, but smaller pairs rely entirely on the grab of the glue.

Dogs and Glue

The glue for a rubbed joint is the old hot animal hide glue. Fresh glue starts with covering the transparent amber granules with cold water and letting them soak overnight. The glue will expand and soften but not become liquid until heated in a double boiler glue pot. Add water or let water evaporate until the hot glue flows from the brush in a thin stream.

Cold wood makes the glue set before it can penetrate the wood, so the pieces should be warmed and ready to go. Brush the glue on both surfaces. Fold them together and rub the two pieces together lengthwise until the glue starts to grab. Stop rubbing when the two pieces are perfectly aligned. Set the joined pieces carefully aside, resting against two battens leaning against the wall. Carefully wipe off any excess glue with a warm, damp cloth or wait a few minutes for the glue to become leathery enough for you to peel it off.

If you use clamps, remember that they will spring broad work into a curve if applied from just one side. Place bar clamps against alternate faces so this effect cancels itself out. Clamps should stay in place for at least six hours, but overnight is better.

If you don't have enough clamps to leave them in place for so long, tap some pinch dogs into the ends of the boards. These dogs look like giant staples, about two inches broad with the spikes tapering on the inner faces. Driven into the end grain of adjoining boards, spanning a joint, the tapered spikes force the pieces together.

Anton Chekhov wrote of another carpenter's dog, one named Kashtanka, who reveled in "the glorious smell of glue, varnish and shavings." Glue does have a tolerable smell—until it goes bad. If glue starts to smell as if a dog wouldn't touch it, it is far past being useful. Glue in the pot gradually loses its holding power, so clean the pot and make a fresh batch every few days.

Coopered Joints

Coopers rightly frown if you call them barrel makers. Staved containers come in all sizes, from firkins to hogsheads, with barrels somewhere in between. Coopers have the longest plane of all (as they are quick to tell you). The five- or six-foot-long cooper's jointer inverts the relationship between work piece and plane. The cutting edge of the cooper's jointer faces upward, with one end of the plane resting on the floor while the other end is elevated on two legs. The cooper stands beside his inclined plane and pushes the stave down it. The cooper works by eye, holding each stave at the precise bevel that, when repeated on the other staves, adds up to a perfect cask.

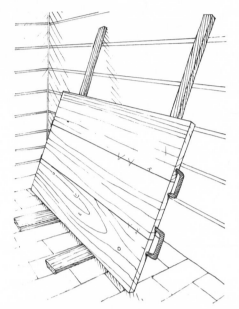

Pinch dogs will let you remove your clamps and use them on another job.

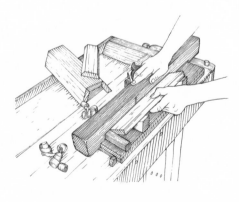

Angled blocks added to the shooting board help you plane consistent bevels for coopered joints.

In one sense, the cooper has it easier than the rest of us. The angle of the edge of each stave is always the same—90 degrees. Before taking a stave to the jointer, the cooper first backs it with the drawknife to the contour of the cask. The back of the stave is a segment of a circle, and the correct angle for the edge is perpendicular to a tangent of that circle—always 90 degrees. No matter how wide the stave, eyeballing a right angle off the edge of the circle gets you a perfect stave. Repeat this a few thousand times and you can get very good at it.

Coopered work for the joiner includes items like porch columns and rounded tops for chests. The process for joiners is the inverse of that used by coopers. You start with rectangular stock, bevel the edges, glue it up, and then round the outside. The bevel of each edge is half the angle between the faces of the stock.

You can always make a full-sized drawing on paper and take the angles from that, but if you're working mathematically, divide the 360-degree circle by as many staves or facets as you will use. With 8 staves, this would come to 45 degrees, so each stave has to cant 45 degrees off from a straight line (180 degrees) to turn its part of the full circle. We're concerned about the inside angle, however, so this is 180 degrees minus 45 degrees, or 135 degrees. Half of that is 67.5 degrees. Any 8 staves of equal width, then, with edges beveled at 67.5 degrees will form a complete circle, or, rather, an octagon that can be planed into a circle.

Once you know the angles, you have to plane them accurately and consistently down all edges. The try square gives you a fixed 90 degrees, but the sliding bevel can be set and locked at any angle you desire. Set at 54 degrees for a five-sided object, the sliding bevel will test the edge angles as you plane the staves. The width of all staves needs to be equal as well, so you can't just keep planing if you make the angle too steep. Often, you plane away too much because you can't see when to stop. A pencil squiggle down the edge will help you judge your progress and tell you when to quit.

Sliding bevels (not all bevels slide, but I say sliding bevel when the tool might be confused with the shape) are especially useful when you're making a piece to custom fit an existing angle. In a hopper-sided tool tote, for example, you can join the sides at whatever angle looks good. With the sliding bevel, record the resultant angle and then use it to test your planing for the inset bottom board. You have no idea what the angles are, but it all fits perfectly.

For work that you are going to do more than once, it's better to use the sliding bevel to make an accurate planing guide and use that for all the staves. A regular shooting board holds wood parallel to a ramp where the side of a plane can ride. Like shooting end grain in the bench hook, this puts the edge of the plane at 90 degrees to the face of the wood and ensures that its edge is square. If you want the edge at less than a 90-degree angle, say 54 degrees, then add little 36-degree ramps to tilt the staves down into the plane. The angle left on the staves will be 54 degrees, and your five-sided column will join the pantheon of pentagonal paragons.

Once set, this sliding bevel can guide your planing to fit a tight bottom for this tool tote.

This is another place for the rubbed glue joint. Rather than trying to glue up the whole thing, work on pairs, allowing each pair to dry before setting up the next. For longer work, clamping is not possible without gluing on temporary blocks to give the clamps a place to grab. Instead, make cradles of two stops nailed to a board, spaced so the staves can sit within them, tentlike. If the spacing of the stops is just right, the joint can be rubbed while sitting between them and weighted until the glue sets up.

Doweled Joints

You can't rub a doweled joint, but that's somewhat the point. The dowels add both alignment and moderate strength. First, however, the holes for the dowels have to be aligned in the two sides of the joint. Here are two methods to try. First is the brad-with-the-head-snipped-off method. Wherever you want the dowels to go in one piece, tack in a brad. Don't drive it in much more than 1/4 inch. A brad is just a headless wire nail less than an inch long, so you can easily cut it with a pair of nippers. Nip off the brads so that only about 1/16 of an inch remains above the surface. Align the two boards and push them together so the brads in one will leave impressions on the other. Separate the boards, pull out the brads by gripping them gently with the nipper, and you have the marks to start your brace and bit.

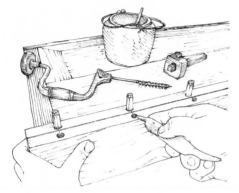

Trim around the edges of the holes in doweled joints before gluing.

The snipped brad method is wasteful but handy. The standard method uses a try square and gauge. Clamp the two boards upright in the vise, face side to face side, the edges to receive dowels upward. Mark across both pieces with the try square at each dowel location, but only with a small mark across the crack between the boards. Separate the boards and, with the stock of the try square pressed against the face side, extend these lines across the edges. Now set the gauge for half the thickness of a board and, gauging from the face side as always, mark this line across the previous ones.

With your dowel centers so carefully marked, take care that you bore the holes straight and deep enough to give a slight clearance at the bottom. The holes, and the dowels, of course, should be about a third of the thickness of the wood. Any wood upthrust by the auger will keep the joint from closing, so cut away a little extra countersink around the top of each hole with a knife. Each dowel needs a saw cut down the side to let trapped air escape, so hold a saw held upside down and drag the dowel along it — just like a cooper's jointer. Check the dry fit of the pieces before gluing and clamping.

Free Tenons and Butterflies

Doweled edge joints get little respect. They keep company with clunky patented doweling jigs and cheap construction. The pegged free tenon, however is the joint that launched a thousand ships. Free tenons not only join the planks of the hull, they also make it watertight. To understand this, look at the joint and think how wood swells when wet, getting fatter but not longer. When

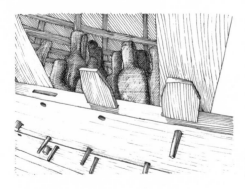

The shipbuilder's pegged, free tenon hides within the planks.

Odysseus had to build a new boat, the radiant goddess Calypso "brought him augers, and he shaped the planks to fit one another and bored mortises in them all. Then he drove the ship together with tenons and pegs."

The pegged free tenon still courses the Chesapeake in the sailing canoes hewn of seven logs. To find a place far from the sea, carry this joint on your shoulder and walk inland until someone asks you what it is. This is not a joint of the carpenter or joiner, but more than any other, it uses wood—as wood.

Norse and Egyptian boat builders used dovetail keys to the same effect as the pegged free tenon. Asian woodcraft celebrates butterfly keys as objects of beauty, exposing them where they join planks with their long grain strength. Europeans use them too, but for some reason have felt compelled to hide their dovetail keys on the underside of the tabletops they hold together.

In work where regularity of appearance matters, snip a template of tin to lay out the butterflies. A few punches with a sharp nail will raise barbs on the far face of the tin to keep it from slipping as you trace around it. Clamp the adjoining boards tightly before you trace around the key to lay out the socket. You can cut the socket entirely with a chisel, but a router plane makes gauging the depth easy.

The dovetail key becomes part of the aesthetic as it spans the joint on the surface.

Rabbet

Dovetail keys are most often inset across a single joint, but in a near-thousand-year-old door at Westminster Abbey, long butterfly keys each span four joints across five oak boards. These five Saxon-era boards are also joined with overlapped shoulders, or rabbets, down their adjoining edges. Coupled with a board rabbeted from the opposite face, the two overlap without adding thickness. This gives greater glue surface as well as making a better seal in unglued joints.

You can cut a rabbet with a chisel, but it's a job made for a plane. Rabbet planes differ from bench planes in that the iron extends the full width of the body, and the shavings escape from the side instead of a central opening. Metal-bodied rabbets generally carry their irons squarely across the body like bench planes, but most wooden rabbet planes are fitted with a skewed iron to help roll the shavings out, draw the plane into the shoulder, and help make a smooth shearing cut when working across the grain.

Skewed or square, a rabbet plane must have its iron set to protrude slightly from the working side of the plane so that it can cut flush on the sides. If the body is even slightly wider than the iron, each pass will shove the plane farther outward, leaving a sloping and terraced shoulder wall.

A simple rabbet plane has no fence to guide it parallel to the established edge; you rely instead on your left hand hanging under the sole to hold it at a constant distance. First, you define the limits of the rabbet precisely with the marking gauge. Then you establish a shoulder somewhat back from the lines, using your fingers as the fence. Next, come back with the plane laid on its side and work with pull strokes until you cut up to the line to your right. Turn the

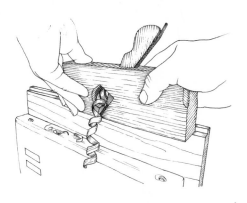

Rabbets are stepped shoulders, easily cut with the simple rabbet plane . . .

. . . or with the moving fillister plane equipped with a fence, a depth stop, and a nicker for cross-grain shouldering.

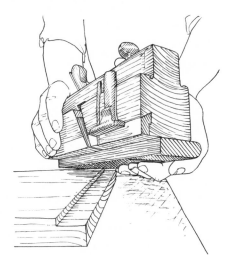

plane back to the original position and finish planing down to the lower line, using the right-hand shoulder wall as the guide.

Fillister Plane

What if the rabbet plane did have a fence that could be set to control the width of the cut? What if it had a depth stop as well, and perhaps a little vertical blade to sever cross-grain before the skewed iron shears it out? This is the moving fillister, and many woodworkers get more use out of it than out of almost any other plane.

The fence on the moving fillister is held to the sole of the plane by two screws and will reveal only as much of the iron as needed for the width of the cut. The depth stop is always handy, but it's wiser to work to gauge lines rather than rely on stops. A gauge can always fence in from the face side, but a stop works only on the side you're planing—and that may be the back side.

The fillister can cut a rabbet down the edge of a board, and across its end as well. Across-the-grain cutting requires two kinds of blades, one vertical and one horizontal. Auger bits have downward cutting spurs to make a smooth cut, and the big crosscut saw has paired vertical cutting teeth followed by rakers. For cross-grain rabbets, you could score the cross-grain with a cutting gauge, a gauge fitted with a blade instead of a scribing point. A straightedge and knife would do as well, but the fillister makes it easy. The fillister plane has a little vertical knife blade called the spur or nicker set just ahead of the plane iron, and just a hair deeper and farther onto the wood. The nicker cuts first, severing cross-grain fibers just deep enough so the plane iron can shear them out without tearing. Just make the first stroke backward across the far corner to ensure that the nicker is deep enough, and then plane away.

Match Boards and Spline

A glued tongue and groove joint is not much stronger than a plain glue joint. The real virtue of a tongue and groove comes when the wood is free to move apart. Tongued and grooved boards can shrink in dry weather, but never show a gap. In tongue and groove flooring, the boards gain stiffness from mutual support—a footfall on one plank has to deflect all the adjoining ones as well.

The old name for tongued and grooved stock is match boarding, and the paired (and sometimes combined) planes for cutting the two elements are called match planes. Where a grooving plane cuts a groove, a tonguing plane leaves the tongue. Both planes need a fence pressed well against the face side of the boards. The size of the tongue and groove is usually reckoned at around one-third of the thickness of the wood. Since the irons of match planes can cut only at a fixed width, the fence is usually fixed as well. Thus, a set of match planes for 3/4-inch boards will have irons to cut 1/4-inch tongues and grooves, inset 1/4 inch from the fence.

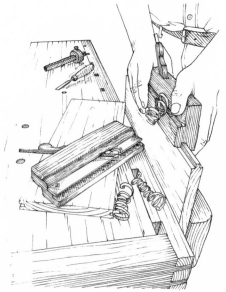

Match planes are dedicated tongue and groove specialists.

Since match planes are intended to work with the grain down the length of the boards, they don't need cross-grain nickers. They do need depth stops, and for these to work, the edges of the boards must be well jointed. One board of the pair retains its jointed edge, and just gets a groove down the middle. The tongued board, however, gets a new jointed edge on both sides of the tongue. You either work to a gauged line or rely on the depth stop of the tonguing plane to make these new edges precisely parallel to the old jointed edge that remains at the end of the tongue.

The new edges on either side of the tongue should come out parallel to the original edge, but they may not be the same depth. Sometimes the off-side element of the tonguing iron is left deeper than the near side. This deeper cut leaves an intentional gap on the back side of a tongue and groove joint to ensure that the face side closes first. Fencing the plane against the face side ensures that this gap is where it should be. If you don't want this gap, a few minutes with the tonguing iron on the stone will make it go away.

When planing match boards, start at the far end and work a section until the plane almost bottoms out, step backward, and do the next section until you have done the whole length. Finish with long walks down the board until the planes will cut no more. The final step with any such joint is a sort of countersinking down the top margins of the groove. Tip the corner of a small, fine-set plane into the groove and run it down the length of the plank so it can lightly sink the shoulders. Repeat this with a pull cut down the other shoulder. The object is to add a little clearance in the hidden middle of the joint to ensure that the visible outside closes tight.

The splined joint uses grooves in both boards so they retain their full width, and spans the grooves with a separate tongue, or spline. A splined joint is a good choice for bench tops built up from narrow, thick pieces. You don't lose any of the precious width, as you would with a tongue.

Because the spline can be cross-grain wood, it can be thinner than the one-third of the thickness required for a tongue and groove. The spline can also make a stronger joint if it is tougher stuff than the boards it's joining. The proper spline has its grain running perpendicular to the joint, and it has to be made in short sections. Plane a wide board of hard wood to the thickness required and then saw spline sections off the end. Again, ease the area on the margins of the grooves with a single pass of a fine-set plane.

Butt Joints, Miters, and Backsaws

So much for parallel, edge-to-edge joints — now we're going perpendicular. In the joiner's alphabet, we're done with pushing one I against another I; now we're making Ls and Ts.

Pushing one timber against another and nailing them together is fastenery, not joinery. But if you cut a shoulder, a rabbet, on the end of one piece and drop the other into it, and then nail it, that's something else. The rabbet secures the wood in one direction and allows the nails to come in from two directions.

A shoulder will let you cross-nail your butted, end-to-end joints.

The miter joints of picture frames and such can also be cross-nailed. Additionally, miter joints in picture frames possess a great aesthetic virtue—the cross section of one piece flows perfectly into the cross section of the other. A molding, no matter how complex, can turn the angle intact and "seem one intire Piece."

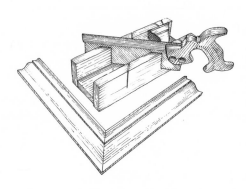

Miter joints allow the molding contours to turn a corner.

Miter joints, like coopered joints, divide the angle of a turn equally between the two joining surfaces. With western architecture based firmly on 90-degree turns, guides for cutting 45-degree miters abound. A miter square has its blade set to form 45 to one side and 135 degrees to the other. Of course if the stuff is large enough, you can position a framing square at 45 degrees to an edge by positioning it so that it crosses the edge at equal numbers on each arm. On smaller stuff, you can use the same method to adjust a sliding bevel to 45 degrees and use it as the bevel square. On the end of a board, a common marking gauge can define a 45-degree angle. Set the gauge to the thickness of the board and use it to run a line parallel to the end grain. Connect the corner on one face and the line on the other and you have 45 degrees.

All of these will give you a guideline for your saw, but a miter box guides the saw itself. The old miter box was hardly a box at all, more like a bench hook with a thick back block with slots to guide the saw. This is now called a miter block to distinguish it from the familiar wooden U-shaped trough of three boards. The front board of the miter box is usually left a bit long to catch on the end of the bench, similarly to a bench hook. This lets you push the work into the far corner of the miter box with one hand and saw with the other.

Miters dwell in the world of appearances, and anything that damages the face side counts against us. Saws leave a rough exit, so you always hold the work in the miter box so that the saw teeth cut into the face, leaving the raggedy edge on the back side. This means you can't reverse a piece in the miter box to cut a reverse angle. Thus, you need a left- and right-hand version of any angle, but many angles means many slots, each one weakening the walls of the miter box. Still, it's easy to make a fresh box or spring for a metal miter box with adjustable tracks for a big backsaw. Metal miter boxes lack a certain aesthetic collectibility, so you might be able to acquire one at a price that is a bargain weighed against its utility.

Metal miter boxes are usually accompanied by their giant backsaws. The back on a saw permits the blade to be made thinner, with finer teeth, leaving a finer kerf than would otherwise be possible. Backsaws are often known by the particular joint they are intended to cut. The smallest and finest are dovetail saws, fine toothed and often with no set at all in the teeth. Next come backsaws in graduated sizes (carcase, sash, and tenon), all of which are used to cut a tenon's shoulders and cheeks. Finally, there are long backsaws intended for use with miter boxes.

The narrow, even, straight cut of the backsaw allows you to cut a miter so true it can't be improved by planing. Much, then, depends on the accuracy of the guide. If, however, you could cut both pieces at the same time while they are held at right angles to one another, any angular error with the saw would be self-correcting.

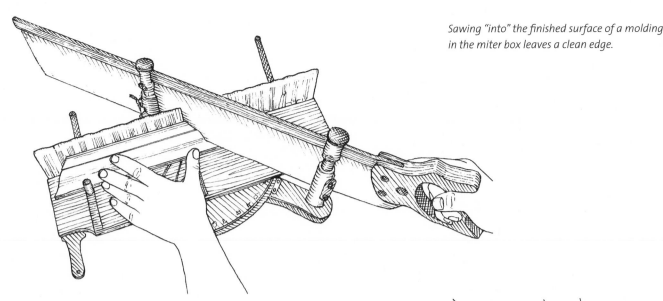

Sawing "into" the finished surface of a molding in the miter box leaves a clean edge.

You can do this in a rough-and-ready way by setting one piece over the other and sawing down through both pieces. The refined version of this is called "kerfing in," a technique that will arise again in mortise and tenon work. In miter work, kerfing in uses a set frame that tightly holds the two pieces at right angles and guides your saw at 45 degrees, right into the miter. You cut both pieces at the same time, not superimposed, but butted with the best miter you can muster. As the blade re-cuts the miter, one side of the kerf cuts one board and the other side of the kerf cuts the other. Tapped together, the two faces close the gap of the saw kerf, making a perfect miter.

Shooting Board, Donkey's Ear, and Miter Jack

A miter box holds the saw at an angle to the work. A shooting board holds your work at an angle to the plane. Just as the saw must work into the molding, so must the plane. That makes the miter shooting board most helpful in shaping small miters that turn an inside corner. The faces of the angled block hold the wood at the correct angle and support it as the plane rides by in its track.

For broad pieces and external miters, you need a donkey's ear. This version of the shooting board holds the work flopped down at a 45-degree incline to the path of the plane. You can work on both sides of the guide, pushing or pulling the plane, so that you can always plane into the molding.

Planing and sawing into the molding prevents splintering and the equivalent of the feather edge we get when sharpening. The feather edge is caused by that last little bit of wood flexing away from the plane rather than staying put and getting cut. This can be a problem with external miters, such as the cornice around the top of a cabinet. You can easily saw into the molding, but planing into it would put you at the worst possible angle for digging in. The donkey's ear allows you to plane across the slope of the miter, not up it (feather edge problems) or down it (digging in problems).

For large mitered pieces, you need a miter jack. The miter jack clamps wood

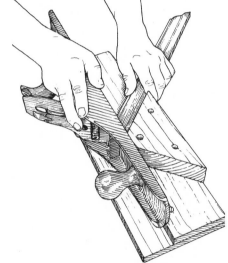

A miter shooting board takes the guesswork out of planing smaller pieces.

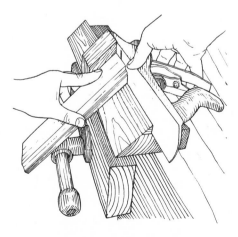

Tall mitered pieces need a donkey's ear.

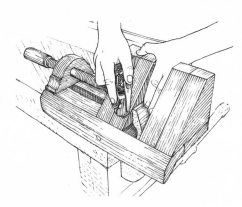

A miter jack lets you plane large, "sprung" moldings.

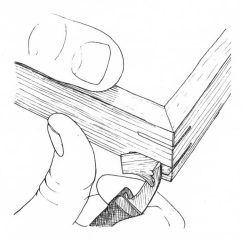

Thin splines—glued into kerfs and trimmed flush—strengthen this miter joint.

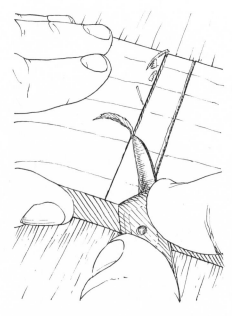

Dados begin with knife cuts to sever the grain and guide the saw.

within its broad beveled jaws that guide the plane or chisel at 45 degrees. This gives you the freedom to come at your molding from any angle. These guiding faces should theoretically be protected by sacrificial thin wood or card stock glued to the wooden faces, but more often you see that the user just tries to stop before cutting the faces with the plane.

If you invest the time to make an accurate donkey's ear, you can make withdrawals for many decades. When you have to work freehand, however, try borrowing the technique of the coopers. Mount a fine set plane upside down in your vise at an incline, just like a cooper's jointer. Moxon speaks of holding the strike block plane upside down in one hand and then thrusting the end of a piece across it. With the plane mounted in the vise, you can work the same way, but now you're free to use both hands to guide the wood down the plane. You can put more of your body into the work, and you always have a good view of any guidelines.

Like rubbed, edge-to-edge joints, miters have no holding power on their own. Unlike edge joints, however, they can rarely be held by glue alone. The joining grain in a miter is halfway between long grain and end grain, so glue does work a bit better than it does on pure end grain. Usually, though, you fasten the mitered pieces to a background or try to strengthen the miter by spanning it with nails, dowels, splines, or keys. The miter can also be just the visible component of a more complex joint. Half-lap and bridle joints can be mitered on just the outward-facing elements, with the other elements overlapped behind them. This is a handy strategy when you want to carry a molding around a corner and still retain mechanical strength.

Housed Shelf Joints

The beauty of the miter is in its continuity of form. It is still, however, a butt joint, and cannot stand on its own. With housed joints, we are back to connecting wood with interlocks. The most familiar housed joint is at the end of a bookshelf, where the horizontal shelf enters a dado, a cross-grain trench cut to receive it in the upright sides. In the simplest version, the end of the shelf spans the full width of the upright, resting in a full-width dado. As with any cross-grain work, the first cut should be made with a knife. Set the knife at one edge of the dado and slide the square up to it. Pull back and slice a deep line across the grain. If you use a paring chisel for this, let the bevel rub against the edge of the try square.

This scoring is not just for marking; it is the first cut of the shoulder. Make a second cut down the line inclined to the waste side. This makes a little V-cut with a vertical wall toward the finished edge, not only severing the cross-grain for a clean shoulder, but also giving the saw a place to ride.

The saw for the cross-grain cut must be straight from end to end so you can be sure of reaching the bottom all the way across. Walter Rose in *The Village Carpenter* recalled how all the handsaws in the shop were sharpened "roach-backed"—slightly convex from toe to heel to allow better bottoming in trench work.

All of this works very well if the dado crosses from one edge of the board to the other. Often though, you want to stop the housing short of the full width of the board. Because the saw is then unable to carry through and clear the sawdust, you have to bore and chisel a short trench at the far end of the cut. This gives a little clearance to work the saw in short strokes. The end of the saw blade is very likely to bump the end of the trench, so leave some extra wood as a bumper until the sawing is complete.

After the knife and saw define the sides, the chisel removes the waste. Start chiseling in from either end to be sure you don't split off too much. Begin the roughing in with the chisel held bevel down, and then finish by sliding the chisel flat across the grain like the finishing strokes of a broadaxe.

Staircase Saw and Router Plane

The staircase saw cuts the housings in the stringers to take the treads and risers. The wooden backs of these saws stiffen the blades, act as a depth stop, and give you handles to hold. The exposure of the blade is adjustable by means of screws passing through the wooden stock into slots in the blade. Set the blade exposure to the desired depth, saw the sides of the housing until it bottoms out, and then remove the waste with a chisel.

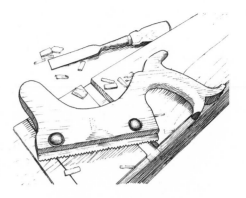

The staircase saw exposes only enough of its blade to cut the dado to the desired depth.

The router really shows how the plane is a guided chisel. With a wood or iron body that rides on the existing surface, the router plane holds its blade extended below to shear a new level surface beneath the first. Wooden-bodied versions, using a stout iron from a plow plane were, even in the tool maker's catalogs, unkindly referred to as an "old woman's tooth."

Once you have sawn the sides of the dado and roughed out the trench with a chisel, set the router plane's iron with just enough depth to take the lightest trim of the rough bottom. Start the router from the outside edges and push (or pull) into the cut. Set the iron a little deeper and go over the whole trench again. Taking the wood down with many passes and resettings of the iron will give you a smooth bottom. Of all the planes, the router gives you the clearest view of the edge at work. Watch the wood and work with a kind of rocking, sawing motion, sweeping the edge diagonally through the grain.

Clear out the waste with a chisel and level the bottom with a router plane.

Dado Plane and Side Rabbet

The dado plane does the work of the striking knife, the staircase saw, the chisel, and the router. You find them in fixed widths, the most common being 7/8 inch, the thickness of quality shelving boards. Dado planes are beautifully designed in the familiar pattern for cross-grain trenching. The scoring knives on both sides of the body sever the cross-grain just ahead of the skewed cutting iron. A depth stop halts the action.

For obvious reasons, dado planes lack a fence, so tack or clamp a batten across your board as a guide. Make the first cut by drawing the plane back across the surface so the nickers get first whack. You can still break out wood on the far end of the cut, so work from the face edge to the back edge. Grab a chisel and

The dado plane does it all, but cuts only a given width.

If the dado is too narrow, a side rabbet plane can ease it open.

The chiseled pocket at the end of the stopped, sliding dovetail provides clearance for the saw.

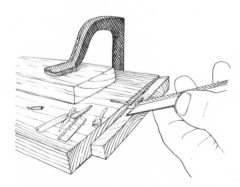

Careful paring makes the tapered, sliding dovetail fit tight only with the final taps of the mallet.

make two deep cross-grain incisions to define the exit—just to be safe. Of course there may not be an exit. The dado plane is not much good in housings that stop short. Here, you're back to the saw, chisel, and router plane.

Technically a dado housing is a mortise, but its orientation to the grain is more like a tenon with cross-grain cheeks and end grain shoulders. On tenons, we can reach the end grain of the shoulder with a rabbet or special shoulder plane. You can't do this in a dado housing unless you have a plane that fits down in the dado and cuts on its side rather than its bottom—a side rabbet plane.

Actually, though, when working down the grain, you need two side rabbets. If the grain is running against you, a side rabbet can't turn around and cut from the other direction, so wooden side rabbets come in right- and left-handed pairs. Metal side rabbets also come in pairs, and in two-in-one versions. If the shelf doesn't fit into the dado on the first go, a few strokes with the side rabbet should clear it up.

Sliding Dovetail

One measure of the security of a joint is how far it must move before it disconnects. A second measure is the direction of disassembly relative to the load. A simple housed dado joint has a grip of less than an inch. The main load is downward, but sides only have to bow out or the shelves bend a little before the books go tumbling down.

The sliding dovetail keeps your books on the shelves by keeping the direction of disassembly perpendicular to the load. It has to be withdrawn six inches or so before it could let go, but even then, the load still pushes the hooked elements together.

The sliding dovetail is found in batten form under plank chairs and table-tops and in the joint between the legs and the pedestal of small tables. Longer sliding dovetails, such as those made to join broad boards in shelving are best made tapered. This allows the joint to fit up easily until the very last, when the wedging action draws the joint tight.

In casework, the top of the shelf is left square and only the lower edge dovetailed and tapered. In other applications, such as making benches or boxes, orient the square shoulder to the weakest end—the end with the least wood. The sliding dovetail is meant to get tight only on the last tap of the mallet, but it can still act as a wedge on that last tap. The square face pushes squarely along the grain, but the dovetailed face pushes and lifts, and can more easily split the board that it is driven into.

Mark lines across the side board the full thickness of the shelf. Set your marking gauge to less than half of the thickness of the side and mark the depth of the housing on the edge; then use the same setting to mark the shoulder of the dovetail on the shelf.

Set your sliding bevel to a one-in-five-inch angle, a little steeper than the usual dovetail, and use it to mark up from the lower corner of the housing on the edge to find the starting place for the long slope. This slope is perhaps a quarter inch in eight. Strike it and the upper, square shoulder with a chisel or marking knife.

Before you start sawing, use the sliding bevel to make an angle guide for your saw from a block of scrap wood. You can cut the sliding dovetail socket with saws and chisels, just as you did the dado. If the sliding dovetail is through, and not stopped, a side rabbet can help you smooth the undercut.

For the dovetail on the shelf, first plane the taper across the grain with a fillister. You can then saw and chisel the dovetail using a guide block cut at the reciprocal angle left on the sliding bevel. Continental European joiners use this joint a lot, and have a special dovetail plane shaped like a rabbet plane with an angled sole. Not by coincidence, the sliding dovetail joint is much like the full dovetail log joint used by Moravian builders. The log joint doesn't slide, but it slopes in two directions at once, and it looks a lot harder to cut than it really is.

Mortise and Tenon for Leg and Rail

The heart of joinery, like that of carpentry, is the proper mortise and tenon joint. It makes our tables and chairs, our doors and windows, with mechanical and aesthetic refinements for each application. We'll take them in turn, starting with the table.

At each corner of a chair or table stands a post. When we make doors and windows, we call these posts stiles, but since we're starting with tables, they're legs. We'll cut mortises in the legs to receive the tenons from the horizontal pieces. These horizontal pieces are called rails, a word that spans chairs, tables, doors, and windows.

We're not joining equally sized pieces, so the thickness guideline of more than a third and less than half doesn't apply. Moreover, in a table leg, two rails intersect at right angles within the leg, the mortises intersecting as well. This makes the tenons each lose some depth to the other. That's why the mortises in the leg are shifted farther toward the outside of the leg, rather than being centered—to give the tenons a deeper reach and equalize the strength of the mortise. You can readily judge the right size and placement for the joint by drawing various options on paper. You know what size mortising chisels you own, and you know the sizes of the wood. Draw a cross section of a leg and see how strong various widths and placements look.

The four legs of the table are alike in size and placement of the joints. Indeed, the more alike you can make them, the better, so start by laying them out all with the same measurements made at the same time. Clamp them all together, face side up on the bench top, square across, and mark them all with the same strokes of the knife or pencil.

Much of the stress and shock on the end of the mortise will come while you're cutting it and test fitting the tenon. That's why it's customary initially to cut the legs, or any piece getting a mortise near its end, about an inch longer than it needs to be. The extra wood helps strengthen the leg against splitting.

Short single dovetails join drawer rails to a table leg . . .

. . . and long single dovetails will connect the legs to this candlestand pedestal.

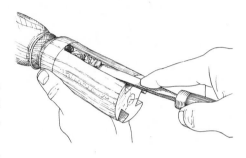

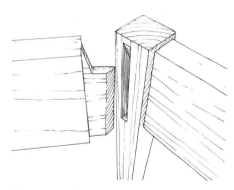

The beveled ends of the rail tenons meet within the intersecting mortises of a table leg.

Unless you want the faces of the rails to sit flush with the legs, reset the mortising gauge after laying out the legs.

After you have assembled the piece and the glue has set, you can trim these protruding "horns" flush to the rails.

The length of the horns may vary, so align the bottoms of the legs and measure up from there to find and mark the height of the legs. Square this line across all four pieces and measure down from it to mark the setback and the width of the rails, again squaring the lines across. The setback keeps the mortise closed and the joint strong, but the greater the setback in the leg, the more must be taken away from the tenon. One-third the width of the rail is the maximum setback I'd consider. Separate the legs and check the square of the lines on each piece, and then carry the lines square around.

The beam of the mortising gauge has two teeth, one fixed in the beam and one movable. Set these teeth to match the width of your chisel and adjust the beam in the fence to place these teeth where you want them on the leg. Ride the fence of the mortising gauge on the faces of the leg to leave parallel lines for the mortises on the back sides. The faces may be the prettiest, but the backs take the joints, and must be true and square.

The legs are now laid out for their mortises, but don't start cutting yet. First, lay out and cut the tenons on the rails. (The tenons-first habit may save you from a potential error later on when making panel doors.) On rectangular tables, do the gang layout for rails in opposite pairs, face edge up. Separate the pieces and check and square the lines all around. If you use a fine pencil at this stage, each line will get checked again when it's scored for sawing.

Set the rail in the bench hook with the tenon to your right, face edge toward you. Stick your striking knife or chisel into the line at the corner of the face and edge, slide the try square up to touch the knife, and hold it there. Now strike the line with the knife across the face, followed by a second knife cut to make the V for the saw.

To score the back side, flip the face edge away from you and place the knife in the far corner of the back side and face edge. Slide the try square up to the knife and then hold it, either with your hand, or squeezed between the back stop of the bench hook and the face edge of the wood. Strike the back line. You now have scored lines across the face and back, and penciled or lightly scribed lines across the edges.

You have established where to start the saw for the tenon shoulder—now you have to know where to stop. If you were to use the same setting on the gauge to lay out both the mortises and the tenons, the faces of the rails would be perfectly flush with the legs. If that is what you want, go ahead and scribe the double lines all around the tenon ends. More often, though, for looks, you want the face of the rail to sit at least 1/8 inch back from the face of the leg—enough offset to look deliberate and not like a mistake. In this case, the mortise gauge is still correctly set for the width of the chisel, but the distance from the fence has to change before you scribe the lines.

Set the rail back in the bench hook and start the backsaw in the V-groove. Keep checking the far side and stop sawing when you're supposed to. Now turn the tenon upright and decide whether to saw the cheeks or split and pare them. The broader the cheeks, the easier splitting looks. This is especially true if

Score across the grain and saw the shoulders of the tenons.

you don't have one of the larger sizes of backsaw, and your large ripsaw seems too coarse for the job.

When splitting, work your way back with several splits so you can see how the grain runs. Splitting always needs smoothing and trimming with a plane or a broad paring chisel. Paring chisels are thin bladed and tang handled—made to be pushed, not hit with a mallet. Hold the rail in the bench hook as you slide the paring chisel, bevel up, across the grain, taking shavings until you just touch the lines left by the mortising gauge.

In narrower tenons, you can finish the surface entirely with the saw, using the same touch-but-leave-the-line accuracy. When you clamp the rail in the vise for sawing, lean it away from you so that you can saw diagonally, watching the lines on the side facing you where the saw goes in and on the end grain where the saw comes out. Turn the piece about and saw from the other side. Slow down as you bring the saw around square to meet the shoulder kerf and finish the cut. Save the sawn-off waste piece for wedges and pegs.

As it is now, the mortise setback requires that you saw away the outer portion of the tenon. This leaves the mortise strong, but means that up to a third of the connection between leg and rail is just a butt joint with no resistance to twisting and no glue surface. Light could even shine through it if the leg shrinks back.

Here's where you decide if you want to leave your tenons bare shouldered or haunched. The haunch is a little stub, square or sloped, that you leave above the tenon to fit into a reciprocal opening above the mortise. A regular haunch is square, and always present in door work to fill the groove cut for the panel. The marginally stronger sloped version is called a secret haunch because it diminishes to nothing and can't be seen once everything goes together.

We still have not cut the mortise or sized the tenon. Which do you fit to the other? In door work, part of the tenon is taken away when you plow the grooves for the panels. Cut the mortise first and you may forget or be surprised by the amount of wood removed by this groove and end up with an undersized tenon. But, working the tenons first, any loss to the groove would be immediately revealed. Making the tenon first also permits you to work by superimposing the completed rail tenon on the leg and confirm the position of the mortise. You already have the guidelines on the legs; the superimposition is just a further check.

Align the edges of the rail tenon with the lines on the leg and transfer the setback dimension. Use a rectangular cabinet scraper laid flat on the tenon and pushed against the shoulder as a square to run this line down the tenon. Saw the waste away, leaving a haunch if you wish. Now lay the tenon across the leg with the shoulder pushed flush. Confirm that the lines for the mortise are good and get ready to chop with the chisel—or bore with the auger.

On long, deep, or large mortises, removing much of the waste with an auger is a great help. On smaller, narrower mortises, the auger holes just slow you down. A table leg mortise is right on the line between auger helps and auger hurts. Since we have two intersecting mortises to cut, we'll do both, starting with the auger.

On broad tenons, split back to the line using the chisel as a wedge . . .

. . . then pare the cheeks smooth with the beveled edge of the chisel facing up.

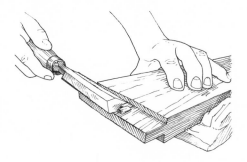

Some like to bore their mortises and pare out the waste.

But chopping with the mortising chisel is usually cleaner and faster.

If we were going to do the whole of the mortise with auger and paring chisel, we could just hold the leg in the vise. But even using the auger, it helps to establish the mortise with a light pass of the mortising chisel. So, set the leg on the bench top under the holdfast, and set your chisel across the grain about 1/8 inch in from the ends of the mortise. Make a light tap with the mallet, driving the mortising chisel in about 1/8 inch. Walk the chisel down the mortise about another 1/8 inch, bevel facing the uncut wood and tap it in again. Work your way along, not going for depth, just making a clean recess to start in.

Find an auger bit smaller than the width of the mortise and set it at one end of the mortise, ready to bore. Set a try square next to the bit and true it up before you begin. A bit of masking tape on the auger or counting the turns will help you gauge the depth. Stop short of the space shared with the intersecting mortise. If you were to undercut this mortise, subsequent chiseling might splinter away the inside corner and weaken the leg.

With the first auger hole complete at one end of the mortise, move to the other end and bore the next one. Let each subsequent auger hole slightly overlap the previous one. You'll need to increase the spacing if a small auger in soft wood keeps getting pulled off track into the overlapping hole. In a perfect world, the final hole takes out the wood between two existing holes and overlaps them equally. Make it a game to adjust the overlaps as you bore along so that this comes out right.

When the boring is done, the mortise is a mess. It's the paring chisel's job to clean it up. With the bevel to the middle of the mortise, start slicing your way back through the webs of wood left by the auger. Until the sides are pared back to the line, the mortising chisel can't do much more than act as a gauge for the width. Were you to drive it in, it could wedge the webs apart and split the leg open. The mortising chisel gets honest work only at the ends of the mortise. Even then, leave that last 1/8 inch until after the other, intersecting mortise is done.

This second mortise is going to be all chisel work. First, for cleaner work, sink a shallow mortise with light cuts the same way we did earlier. The bevel always leads—that is, as you work away from a previously cut area, the bevel faces the uncut area. This orientation causes the chisel to ride back as you drive it down, and forces the chip out behind it.

Set the leg on the bench top under the holdfast. You may also want to put a hand screw clamp around the leg as extra insurance against splitting. Some shops use a special mortising stool, a low, stout sawhorse with a broad top. There you can sit on the work to hold it as you mortise. You can hop up on the workbench and sit sidesaddle on the work just as well.

Most workers like to get to the bottom of the mortise as quickly as possible and work out from there, so that's what we'll do. Position the chisel across the grain in the middle of the mortise with the bevel facing the right. Drive it in. Pull out the chisel and turn it so that the bevel now faces left. Set it less than 1/4 inch away from the back of the previous cut and drive it in again, pushing a chip to the middle. Repeat this left and right beveled chopping, until you have excavated a V to almost the full depth of the mortise. Throughout this, the bevel has been facing the walls of the deepening V and the flat side has faced the middle.

Once the V is to the bottom, it gives the chips a place to go as you work outward to the ends of the mortise. (Starting with a single auger hole and expanding it is another option.) I usually work without a central cut, and march up and down the mortise with a series of deepening cuts. In any case, working with the mortising chisel alone is smooth, precise, and efficient. Soon enough you'll come to the bottom. We have preserved the last 1/8 inch of the ends of the mortise to give us a shoulder to lever against. Now we can turn the chisel so that the flat side leads and work our way back into the ends, squaring them down to the bottom. You can check the squareness of the mortise end by holding the flat of the chisel or the edge of a ruler against it and comparing its inclination with a try square.

If you have left haunches on the tenons, cut the places for them to fit above the mortise. The tenon should make a good push fit. As when chopping the mortise, the leg will resist great pressure along the length of the grain but may be easily split by pressure across it. As you test fit each tenon, reach into the intersecting mortise with a pencil or scratch awl and mark where to bevel off its end. Shear off the bevel with the paring chisel and test the fit of the three pieces together.

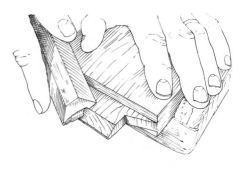

Pare the tenons to a miter. The sloping stub above the tenon is the secret haunch.

Working carefully, everything should fit square and flush. If one of the tenon shoulders shows a gap, you could use your backsaw to kerf it in. Bring the framework together, squared up as desired. Set the backsaw flush against the mortised piece and saw the ill-fitting tenon shoulder, taking it back one even saw kerf. Hold the joint from shifting as you saw the tenon's other shoulders in like fashion. Now all the shoulders of the tenon are at least one even kerf away from the mortised face. Knock the joint in and it should close flush. Repeat if needed. Of course, this kills any fine surface, but there are times and places where it's handy.

Finally, you can drawbore and glue the joints. Positioning the pegs is a matter of judgment best developed from observation. If you set the pegs too far from the shoulder, the leg may shrink back to them. Set them too close to the shoulder, and the leg can split, so give yourself at least a 1/4-inch setback.

In joinery, you always need to be careful about splintering. When you bore the peg hole through the cheeks, stick a scrap in the mortise and on the far side. This will make a clean cut and help the auger run true.

It's easy to mark the tenon for drawboring if you're using a screw-tipped auger. Just stick the tenon in the mortise and push the auger back through the hole in the cheek, leaving the impression of the screw on the tenon. Pull the tenon back out and bore its hole, offset up to a quarter of the peg diameter toward the tenon shoulder. Splintering in the tenon can also weaken it greatly, so back it up with a scrap, and clamp the sides if your auger screw might split it. Prepare the way for the final peg with a tapered iron or hardwood drawbore pin to burnish and pull up the joint between the table leg and rail.

Foxtails and Keyed Tenons

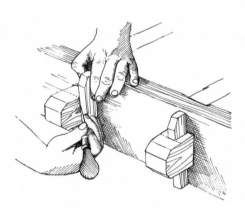

Foxtail tenons take a one-way trip into the undercut, stopped mortise.

The mortise and tenon joint takes endless forms to adapt to every situation. We just finished drawboring a mortise and tenon to lock it in place, but what if we couldn't or didn't want to peg it? If the tenon passes all the way through a piece, as it will in frames and sash, we could expand the tenon with wedges from the far end. If we expand the mortise a little bit on the outside, the wedged tenon will expand to fill it, forming a dovetail.

This works fine for a through tenon, but a blind tenon can also be expanded in similar fashion. A blind wedged tenon, or foxtail, goes into its undercut mortise fitted with wedges that expand the tenon as it bottoms out. Once it's driven in, there's no going back, and if a wedge is too deep or breaks, you're stuck with it. If the wedges expand the tenon enough to make it impossible to withdraw, but not enough to make it tight, you're also stuck. You might go so far as to make a cutaway test version of your foxtail tenon to test the joint. In any case, its a handy thing to show clients.

You can also blind wedge a peg. This is most handy in the odd instances when you need to lock a peg into end grain. In log building, this is how you peg in door casings. Saw a slot in the end of the peg and fit in a wedge. When the peg hits the bottom of the hole bored for it, it expands and locks the peg in place. Saw the peg off flush and drive in a wedge from the outside. Pegs check in, but they don't check out.

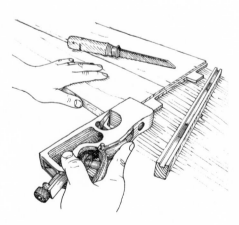

Keyed, through tenons are called tusk tenons in joinery, but they lack the stubby, hidden shoulder of the carpenter's joint.

Keyed tenons are the extroverted opposite of the foxtail joint. In a keyed tenon, the mechanics of the joint are out for everyone to see. Where a foxtail tenon can never come apart, the keyed tenon is easily knocked loose for disassembly. A true tusk tenon is a haunched construction joint that pokes through a timber and bears a tapered key; but nowadays, any keyed tenon is called a tusk tenon. The key is truly more of a wedge and pulls the joint tighter as it is driven in. Fitting is easy if you make the wedge first and trace its taper onto the protruding tenon. Make the mortise for the key a bit further into the side toward the shoulder. This allows the key to pull up the joint as you drive it down.

Double Tenons and Clamp Joints

On pieces needing extra strength, two tenons are stronger than one. A double mortise and tenon joint gives more glue surface while retaining the strength of the mortised piece. Very broad pieces also benefit from double tenons. Say you are making a standing desk with broad rails forming the body. A tenon left the full depth of the rails would be quite broad and strong, but it would have to fit into a very long, slotlike mortise in the leg, weakening it considerably.

The solution is to break the broad tenon into several smaller ones separated by a shallow tongue. This is also the solution for "breadboard" ends on tabletops and lids. Commonly called clamp joints, the tenons and tongues on the ends of the broad table boards fit into mortises and grooves in the narrow end boards, or clamps. The clamps keep the broad boards in line and reduce the exposed end grain in finished pieces.

Trim the end grain of the clamp joint with a low-angle shoulder plane.

You can saw the shoulders of tenons across broad boards in the usual way, but the cheeks must be split out and planed smooth across the grain. Once the tenon stock is the proper thickness, saw out the waste separating the individual tenons with a back saw followed by a turning saw. The groove and mortises in the clamps can be cut by the same method used in undertaking frame and panel work, which comes next.

Panel and Frame

George Eliot's 1859 novel *Adam Bede* begins in a joiners' shop, where a distracted young worker calls out, "There! I've finished my door today, anyhow." To his embarrassment, and to his coworkers' amusement, he is holding up an empty frame for approval, having forgotten to make and fit the panels. Eliot may have been a decent novelist, but this scene is beyond silly. Panel-in-frame construction is the soul of the joiner's trade.

Panel-in-frame construction allows the wood to move.

Defining the perimeter of a door, cabinet, or wainscot with the stable long grain of the wood ensures that it will not swell or shrink in its outer dimensions. The broad panels, freely moving in grooves plowed within this frame can then expand and contract harmlessly. The design enables a dynamic material to retain a set outer form. The result is a door that does its job; it opens and closes at your command, never sticking in the summer, and never opening up cracks in the dry winter.

In the days of the village joiner's shop, making a basic, four-panel door was considered a good day's work. This meant ripping, planing, and grooving the stiles and rails of the frame, laying out and cutting the ten mortise and tenon joints, planing and molding the four panels to fit into the grooves of the frame, and then fitting the whole together. I am also told that just because this was considered a good day's work, it doesn't mean that anyone ever did it.

Plow Plane

Making a molded frame and panel requires a marking gauge, a mortising gauge, a mortising chisel, and three special planes, the ovolo, the plow, and the panel raising plane. There are, however, ways to accomplish the tasks of each of these planes using other tools.

Consider the central tool for a moment. The plow plane makes a groove along the grain of a board. Each plane is fitted with eight or more irons of different widths. Some plow irons are stamped with numbers in a system that makes a 3/16-inch-wide iron a number 2 and a 1/2-inch-wide iron a number 7. This is based on starting with the smallest iron, the 1/8 inch, calling it the number 1 and counting upward in sixteenths. The irons taper greatly in thickness from nose to tail, making a double wedging action with the wooden wedge. This taper is not standard, so take care when buying a plow and irons—there are lots of cats and dogs out there.

Plow plane irons also have a groove down their backs to position them on the iron, skatelike sole. This narrow skate allows you to use irons of differing

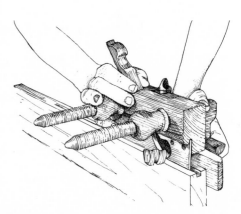

Plow the grooves to hold the panel.

widths, but it limits the smooth working of the plane to the most even-grained wood.

You can't really be knocking on the ends of a plow plane to adjust the exposure of the iron. To make the cut deeper, you tap directly on the top of the iron. To release the iron, you tap harder, enough to release it so you can pull out the wooden wedge. Each iron also has a little shoulder or "sneck" at the top end to let you tap the iron upward. If the iron is too deep, you usually just release it and start over with a too shallow set and gradually make it deeper, tapping alternately on the tops of the iron and the wedge.

The adjustable fence in a plow positions the iron a given distance in from the edge of the board, and the depth stop controls the depth of the cut. The fences of plow planes may be adjustable by wedges or by wooden screws. Like braces, plow planes were, and are, recognized as things of beauty. Presentation plows of ebony, ivory, and brass go to collectors for huge sums. You may not yet have a proper plow, but you can do its job with a double-toothed marking gauge and a chisel. Just run the lines from the marking gauge quite deep and then start shaving out the groove with long, bevel-down strokes of the chisel.

Molded Frame

We'll make a simple molded frame with a single raised panel. I'll call it a door because it's the shortest name, but several of these basic units could make a cabinet—more could wainscot an entire room. The mortise and tenon process from the table leg and rail joint is modified here to allow for the groove and the molding. A name changes too. In frame work, rails are still rails, but the vertical posts are now stiles.

The stiles and rails of this door will have a decorative molding down their inside edges. This molding is a simple quarter-round shape, set down into the corner. This ovolo is easily struck with a single molding plane, a combination of planes, or a scratch stock. The single wooden molding plane is simply called an ovolo. It will cut both the round surface and the two square shoulders, or fillets. Alternatively, you can sink the square shoulders with a moving fillister plane and shape the round with a hollow—a common wooden molding plane.

However you cut the molding, it's safer to work from gauge lines than from stops and fences on the planes. Having said that, you must first set your gauge according to the cut of the ovolo plane. Then, when working with the ovolo plane, use it within lines established by the gauge. It's easy to rock or reverse the plane, but the gauge lines never lie.

The face edges in panel work are turned to the inside of the frame, so the first step is running gauge lines for the ovolo around the face edges and face sides. We'll mold just the face of this door; the back will be plain.

Set the double-toothed mortising gauge for the groove that will hold the panel. The width of the groove must match one of the irons on your plow plane, and for the simplicity of the moment, the width of your mortising chisel. Run this double line all around the inside face edges of the frame, gauging in from the face sides. Continue this double line around the ends of the rails where the

Strike the molding with an ovolo plane, or with a fillister plane followed by a hollow (shown here).

tenons will go. These are going to be through mortises (and through tenons) so run the double line on the back edge of the stiles where the mortise will pass through.

The groove and the ovolo complicate layout somewhat, but only the impact of the ovolo needs attention right off. Earlier, in the table rails, the tenons had level shoulders to fit flush against the flat face of the legs. In this case, however, we have a curvy, molded face to deal with. Cruel as it seems, we're going to cut away most of the curvy ovolo on the stiles where the rails intersect. This means that the shoulders on the face side of the rail must be that much wider than the shoulders on its back.

Lay out the offset shoulders of the rail tenons, marking both pieces in tandem, placing them face sides together, face edges up. The narrower span of shoulders on the back sides equals the total width of the door, minus the full width of the two stiles. The shoulders for the face side are each an ovolo-molding's width farther out. Separate the two rails and run the lines around with the try square.

For the stiles, clamp them together and lay out just the inner and outer dimensions of the frame. We'll get the mortise by superimposing the tenons later on. If you want to mark the mortises now, remember to subtract for both the setback and the loss to the groove in the rail.

Moldings are there for looks, so it's good practice to wait as long as you can before striking them into your stock. The groove, however, is mostly mechanical, so start with it. Mount the stock on the bench top, face edge up, face side to you. Adjust the plow to fit right between the gauged lines and work your way back from the end, bringing down a section at a time. The depth is usually no more than 3/8 inch.

Now saw the offset shoulders on the rails, squaring and scribing a little V for the back saw as before. Don't saw the cheeks yet, as the intact wood will help the ovolo plane ride smoothly to the end, right over the saw cuts for the shoulders. You can use the ovolo plane with your frame pieces either standing face edge up, or lying face side up, but you have to choose one and stick with it—unless you have the world's only symmetrical ovolo plane.

As with many molding planes, ovolos are usually "sprung," that is, they are used with a cant to the side. They still progress straight down with each pass, but you hold them with a constant sideways tilt. Spring helps the shavings pass out of the plane and keeps the curviness of the edge more square to the length of the irons.

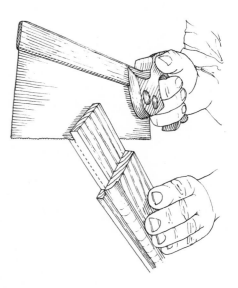

Saw the cheeks of the tenon.

The word for planing a molding into a piece of wood is "sticking." It can be confusing because it sounds like you're putting something on, when you're actually taking something off. You have gauge lines to guide you, and if you want or need to, you could cut to these gauge lines with a moving fillister, and shape only the quarter round with the ovolo. It's good practice to do as much as you can with easily sharpened tools and then use the slower or harder-to-sharpen ones for the finish.

When the ovolo molding is completed, saw off the cheeks of the tenons, taking care not to let the plowed groove pull your saw into it and make the tenon too thin. You may even want to remove the wood on either side of the groove and re-strike the lines with the mortising gauge.

Now saw off the wood for the setback of the tenon, stopping at the face side shoulder. A rail tenon for an unmolded frame is always haunched to fill the groove in the stile. Here, the offset of the shoulders fills that space.

Lay the completed tenon across the stile and confirm the layout for the mortises. These are through mortises, rather than blind mortises such as you find on a table leg. Through mortising is easier, because you can work from both sides and meet in the middle. You can also more easily check the ends of the mortises with the try square, fitting the blade all the way through.

It's standard practice to chop the mortises before plowing the grooves in the stiles, even when they're the same width and placement. See that the tenons pass a test fitting in their mortises before you plow the grooves and "stick" the moldings up to the gauge lines. You'll be plowing right over the open mortises with a risk of damage to the wood if the plow skate hits the far end like a blunt axe. Plow with care.

Scribed Joint

If you push a tenon into a mortise now, you'll see that the ovolo molding on the mortised piece is in the way. We'll now cut away the ovolo on the stile, leaving just enough on the inside corner to make a nice joint with the ovolo on the rail. From the front, this joint will look like a miter, but it won't be. Instead, we'll cut the ovolo on the rail to cup over the ovolo on the stile in a scribed joint. The scribed joint is superior to a miter joint in that it will never open up a visible crack when the wood shrinks.

To rid ourselves of the excess ovolo, we must first determine how much to leave behind. Start by fitting the joint together as far as it will go. Find a point on the tenoned rail about 1/4 inch farther onto the board than the inner limit of its ovolo. Now mark this point on the ovolo of the mortised stile. Pare away the ovolo on the stile from this point to the end, making the surface square and fair.

Now we scribe the ovolo on the rail tenon to fit over the stub of the ovolo on the stile. When a scribed joint is viewed straight from the face, the junction appears to be a simple 45-degree angle, as indeed it is, for mitering one of the intersecting ovolos is the first step in scribing the joint.

Use a bevel gauge or a good eye to pare the end of the rail ovolo to 45 degrees. Better too little than too much. Turn the rail edge up and look right down at the mitered ovolo. You'll see that it looks like a perfect little quarter circle. It is, and that's what you need to take away with a scribing gouge. This gouge has a dead flat convex face and a beveled inner face. These are sold as in-cannel gouges in both the thick firmer and thin paring styles. You can use an out-canneled carving gouge in a pinch, but it's the proper, thin, ultra-sharp scribing gouge that gives you the edge.

Work your way in and down with the gouge, until you come right up to the line left by the mitering. Continue the reach of this hollow until it will fully

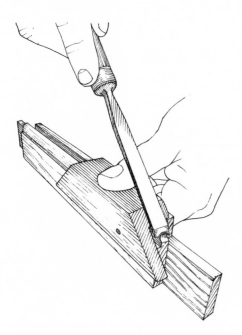

Guided by a miter template, pare the ovolo to a flat miter . . .

. . . then undercut the profile with the scribing gouge.

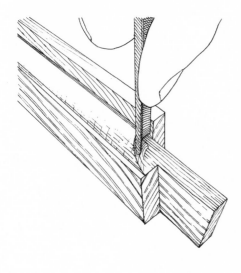

house the stub of the ovolo on the mortised stile. If, when you assemble the joint, the fit of the tenon is a bit snug, it's better to use a bar clamp for slow pressure. Driving the pieces together quickly with a mallet could break open a scribed tenon that could otherwise have been fixed with a little more paring. It's satisfying to watch as this joint closes up—so exciting that we almost forgot the panels.

Panels

Now that you have the assembled frame, you can measure the span between the grooves for the panels. The panels will fill the space within the frame, riding freely in the grooves. By their free expansion and contraction within these grooves, the panels can respond to changes in humidity, without the entire door expanding or shrinking. The entire design is based on the panels changing their width, but what width do you begin with?

This causes much anxiety. What is worse than making a door and having a panel shrink so much it leaves a gap? Before you begin work on the frame, put the panel stock in the driest place you have to shrink the wood to its minimum width. Oak is notorious for shrinking again once a new surface is exposed, so even then, you can't be fully sure until the final planing. Aim for the middle width, fitting the well-seasoned panels so they have room to both swell and shrink. Wider panels need deeper grooves to move in, but even the foot-wide panels in wainscot around a fireplace pull back only about 1/4 inch in the dry heat generated by winter fires.

Panels are too wide for layout with a regular marking gauge. A panel gauge is not only longer in the beam, but wider in the fence as well. The fence also has a rabbet to fit around the face edge of the panel and keep it from rocking over. The panel gauge can also help you after you have sawn and planed the outer dimensions of the panel. Even if you are going to just taper the margins of the panels with a drawknife, you may want guidelines for your free hand work.

Panels just need to fit in the grooves, and a drawknife can do that, but the classic form of the raised panel, with the margins sloping away from the rectangular flat, comes from planes. The planes have to work well across the grain and flush to a shoulder, but that's all. Panel-raising planes come in varied forms, but generally they have a broad, skewed iron bedded at 50 degrees, shaped to cut both the canted margins and the flat tongue to fit into the grooves.

Dedicated panel-raising planes may also have fences, nickers, and depth stops, but other tools can do these jobs as well. A cutting gauge can score the cross-grain, as can a knife guided by a batten clamped across the panel. This batten can then guide your rabbet plane as you first cut down and then cant over to plane the bevels. Whatever planes you use, finish both cross-grained ends first so that any break-out at the end of the strokes will be removed by the long grain planing that follows.

Panel-raising planes are among the aristocracy of bench planes, but they are governed by a humble scrap called the mullet. When this grooved stub, cut off one of the ends of a stile, tells you that the panels will fit in the grooves, you

Lay out the width of the panel with a panel gauge.

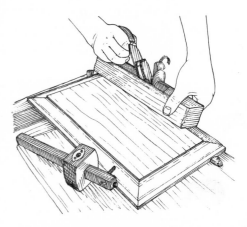

Work the cross-grain ends of the panels before you plane the long-grain sides.

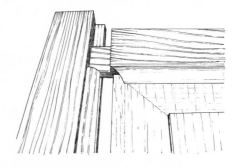

The scribed end of the ovolo cups over the ovolo on the stile.

stop planing. As best I can tell, the name, "mullet," has a Latin derivation from a word for shoe. This makes sense in the way you try the mullet on for size. If the mullet fits—don't plane it.

The 1734 *Builder's Dictionary* warned joiners never to glue or nail their panels: "This will give Liberty to the Board to shrink, and swell without tearing; wheras Moldings that are nailed round the Edge, as the common Way is, do so restrain the Motion of the Wood, that it cannot shrink without tearing." Don't forget the panels, and don't forget to leave them free to move.

Sash

Sash lies deep into joiner's territory. Sash making has long been a specialty trade in woodworking. A person who would think nothing of building his own home would still probably buy his windows readymade. Sash doors are used in furniture as well, and the principles for making them are the same.

Like a panel door, a basic window sash is a rectangular frame of four sticks: two vertical stiles and top and bottom rails. The space within the frame is divided into smaller rectangles by narrower muntins. The frame and the muntin have a shoulder called the glazing rabbet, planed into one side to take the glass

Sash making is a special skill within the joiner's trade.

and the putty. The molding on the inside of the sticks reduces their visual obtrusiveness, but little of their strength.

Strength figures in another aspect of the window. Study a wooden window and you'll probably see that the vertical muntins are continuous and the horizontal ones are in segments. In sash doors, hung from hinges on one side, many choose to leave the horizontal muntins continuous and segment the verticals. In either case, the segmented muntin always gets made as one long piece, molded and tenoned, and gets sawn apart only as the final step before assembly. The most familiar molding for sash is also the inset quarter-round ovolo. I will describe making a four light sash with rails tenoned into the stiles. The upright muntin will be a single long piece, and the cross muntin will be cut into two sections before assembly.

As in panel doors, the molding adds some complexity to the tenons. Rather than two level shoulders, as on a common tenon, the tenons on molded sash sticks have shoulders at three different levels. If you look at the profile of the tenon, you see that it butts against the fillet above the ovolo, all around the ovolo, and against the glazing rabbet.

If you are making more than one window of a given size and pattern, make a pair of guide sticks for laying out both the vertical and horizontal pieces. The vertical guide stick holds measurements for the stiles along one edge and the upright muntins along the other edge. The process, then, is to lay out all the pieces while they are still square and unmolded. Cut all the mortises, but saw just the shoulders of the tenons. Then plane the moldings, finish the tenons, scribe the copings, and assemble the sash.

It starts with the wood. Saw and plane all of your stock to the finished rectangular sections, trying to orient the growth rings of the wood perpendicular to the major surfaces. This minimizes distortion and windows that stick in the summertime. Mark all of your pieces so that you can maintain them in the same up-and-down, left-and-right order in which you laid them out.

Set the points of your double-toothed mortising gauge to the width of the mortising chisel, perhaps 1/4 or 3/8 of an inch. This must also be equal to the width of the fillet or listel (the square shoulder) between the ovolo and the glazing rabbet. Adjust the fence of the mortising gauge according to the molding made by your sash ovolo plane. Gauge all the way around the interior edges of all pieces and on the outsides of the stiles where the mortises will come through.

Now set a single-tooth-marking gauge to the face width of the ovolo molding and run that line all around the faces of all interior edges. Reset the gauge to the width of the glazing rabbet and run that line all around the back faces. Better to have these gauge lines than to rely on the adjustments on your planes.

Mark all the mortise and tenon locations from the guide sticks, leaving extra length for horns on the stiles to strengthen the ends during mortising. Chop all the mortises and then saw only the shoulders of the tenons.

American pattern "stick and rabbet" planes cut both the molding and the glazing rabbet at the same time. British practice uses a sash molding plane to cut the decorative profile, and a sash fillister plane to cut the glazing rabbet. The

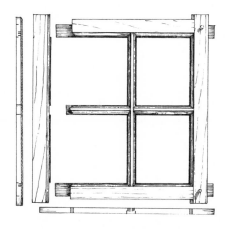

Begin by making guide sticks marked with the mortise lengths and tenon shoulders.

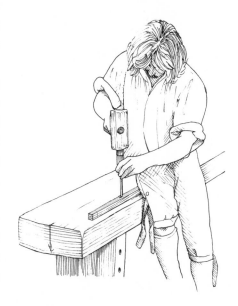

Chop all the mortises and saw all the tenon shoulders.

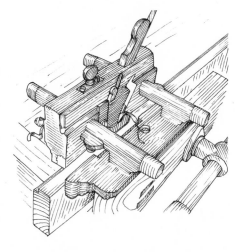

The sash fillister fence rides against the face side as it cuts the glazing rabbet on the far side.

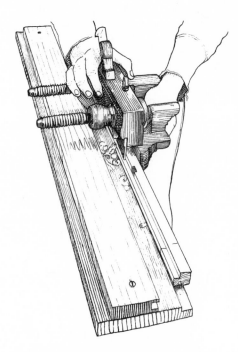

A plow plane can also cut the glazing rabbet. Hold the muntin in the first position of the sticking board.

Move the muntin to the second position on the sticking board and plane the second glazing rabbet.

LEFT: *Set the muntin in the groove of the sticking board and plane the first ovolo to the gauged lines.*

RIGHT: *Flip the muntin and finish the second ovolo.*

sash fillister differs from the ordinary fillister in that its guiding fence mounts on arms like on a plow plane to ride against the face side (the molded side) and lets the cutting body reach over and cut the other side. Fencing from the face side, the sash fillister throws any variation in stock thickness to the outside of the house. A plow plane can also cut the glazing rabbet, as can a plain fillister or a rabbet plane.

The stiles and rails of the sash are stout enough to mount in the vise or on the bench top for planing. First plane the rabbets to the gauge line, then plane the moldings as done earlier when making the panel doors. You will be planing up to the edge of the mortises and through the sawn shoulders of the tenons.

Muntins are too tipsy to just mount on the bench, and need a special cradle, a "sticking board," to hold them as you work. The sticking board has shoulders and grooves along its length to seat the muntin firmly for each of the four steps in planing. The lower shoulder of the sticking board holds the muntin in two positions as you work the glazing rabbet. The groove then holds the muntin as you plane the ovolo on both faces.

When all the pieces have been fully molded, saw the horizontal muntin into two pieces, right between the paired saw cuts made for the tenon shoulders. Keep track of their order and orientation so that they can go back together the same way. All that remains is to finish the tenons and assemble the sash. On the rabbeted sides you can just saw or split away the waste from the cheeks. On the molded sides, though, you scribe away enough of the ovolos on the tenoned pieces to fit exactly over the ovolos on the mortised pieces.

A miter begins the scribe. Using a template or a good eye to guide the paring chisel, slice off the corners of the tenon ovolos at 45 degrees. Now turn these mitered tenons sideways and undercut them by pushing in from both sides with a scribing gouge. On the broad tenons of the rails, you can slice off most of the molding on the mortised piece, leaving only a half inch or so of the ovolo toward the inside. The ovolo on the tenoned rail or muntin gets scribed back only as much as is necessary to house the little piece left on the mortise face—the same technique used earlier in making the frame and panel. Sash goes together

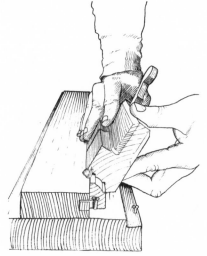

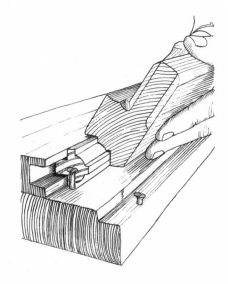

in much the same way as a panel door. There are more pieces to keep track of, but guide sticks and consistent moldings will make the work easier.

Easier has always been the object of the American inventor. The stick and rabbet plane combines the sash molding plane with a rabbet plane so both elements get cut at once. Since it is usually two planes held together by screws, you can insert spacer blocks to adjust the gap between the rabbet and the molding.

Another American device that attempts to stick sash molding into its ditty bag is the combination plane. More commonly known as the Stanley 45 or 55, these nickel-plated assemblies are more like a Civil War surgeon's kit than a plane. They come with a passel of interchangeable irons, including sash-molding, tonguing, grooving, beading, reeding and so forth. Because it uses skates instead of a proper sole to hold the wood down until it's cut, it makes a good plow plane, but a mediocre everything else.

Many workers use a 45 regularly, but only for grooving with the grain for mechanical joints. In a pinch, you can grind an iron to match an odd molding profile and mount it in a combination plane. A simple scratch stock, just a scraper filed to the wanted profile mounted in an L-shaped wooden block, will probably do a cleaner, albeit slower job. If you want fine moldings fast, you need molding planes, and there are plenty out there.

Molding Planes

The jumble of molding planes seems bewildering—too complex to comprehend. American practice uses many specialized planes, each doing one complex molding. In the British tradition, the joiner tends to build up complex shapes with a few planes.

I recently bought a chest of molding planes owned by three generations of English joiners, E. F. Margetson, Tom Killner, and Robert Simms, respectively. Chests of tools are time-traveling vessels, and each passenger has a story to tell. There were the usual number of odd planes in the chest, but the main players

Saw apart the tenoned muntins and scribe the joints.

A small set of hollows and rounds.

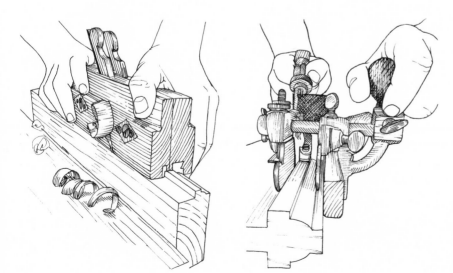

LEFT: *American practice more often uses a "stick and rabbet" plane . . .*

RIGHT: *. . . or a combination plane fitted with a sash iron.*

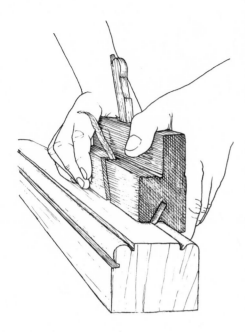

Most corner bead planes have a boxwood wear strip.

were five basic shapes: nine pairs of hollows and rounds, ten side beads, seven ovolos, five ogees, and a few planes for cleaning up. You can easily pick out the most used planes in the chest by the thick, black, almost plastic accretion of sheep's tallow left by the grip of the joiner's hands. English planes are always dark from linseed oil, but some of the hollows and rounds in this set had been used so much that they looked as if they had been dipped in tar.

You reach for the hollows and rounds so often because of their versatile simplicity. Each hollow forms a convex profile on a surface; the matching round forms the concave profile. Used individually, they can form everything from the flutes on columns to drop-leaf table joints. Used in combination, a hollow and round can form complex moldings, particularly the classic wave-shaped ogee.

There are 9 even-numbered pairs of hollows and rounds in the chest. A full set would number 18 pairs, but half sets of 9 pairs are most common. The numbering systems varied from maker to maker, but the smallest planes in this set made by Griffiths of Norwich, England, are stamped with the number 2. They work a 1/4-inch-diameter circle with irons 3/16 inch wide. The largest pair, stamped 18, works a 6-inch-diameter circle with irons 1 1/2 inches wide. All the plane irons are set at middle pitch, 55 degrees, and at a skew of 10 degrees off of perpendicular to the progress of the plane. This skewing made a smoother cut but required extra work and workmanship from the plane maker, adding a 10 percent premium to their cost.

In the signs-of-use contest, the beads run a close second to the hollows and rounds. Among the 10 beading planes in the chest is a matched set of 7 running from 1/8 to 7/8 inch across. These are properly called side beads, because they work on the edge of a piece, cutting a half round isolated by a little square-bottomed valley called a quirk. Worked down both sides of the corner of a timber, the side bead forms a three-quarter round. The shape of the bead is not just decorative in the moment; it also wears better. Sharp corners look fine to start, but on a post or plank, they quickly take dents and splinters. The side bead moves the sharp edge back away from the corner to the quirk, protecting it behind the round and sturdy shoulder of the bead.

The bead planes are themselves protected by inset boxwood strips. Hollows and rounds work only 60 degrees, one-sixth of a circle. Bead planes work the maximum 180 degrees, with a narrow protrusion to cut the quirk side. Only on the cheapest planes is this protrusion a fragile extension of the beech body. On professional-quality planes like these, the maker inserts a diagonal boxwood strip to form this point.

Boxing not only replaces the beech with a harder wood, it also changes the grain direction. The glue that held the boxwood in the 3/4-inch bead plane has failed, and pulling out the strip shows how it was sawn on a bias to its grain. Orienting the grain at 45 degrees to the length of the strip brings it parallel to the plane iron, eliminating short grain behind the iron. The boxing at the nose of the plane is also made stronger by the diagonal grain running back into the body of the plane. Planes may be just blocks of wood with pieces of steel in them, but wisdom waits beneath the grease and dirt.

The ovolo and ogee molding planes in the chest are a conservative lot, all Roman style—based on circles. This makes things simpler, for if this were a late nineteenth-century American joiner's chest filled with the then-popular Grecian-style molding planes, we'd be facing an indescribable variety. Moldings in the Grecian style are based on ellipses and parabolas. Circles come in infinite sizes, but they have only one shape. Ellipses come in infinite shapes, as well as infinite sizes, and half of that infinite variation was made into molding planes.

We used an ovolo plane earlier, making the panel door and the sash. The ogee planes are the next step in complexity, imparting the classic S-curve to the wood, combining it with a little fillet to give it more definition. All the ovolo and ogee planes work on the spring—tilted over as they cut.

Finishing out the molding planes were some clean-up planes. There were snipes-bills to refine tight turns and side rounds to help smooth curves. Most molding planes cannot be turned around and worked from the other direction when the grain of the wood proves troublesome. Clean-up planes come in left- and right-handed pairs so they can work from either direction to go with the flow of the grain.

Planes work as well today as they ever did, but we often forget the roughing in part of the job. Save your planes and your time by using a gouge to rough in a hollow, and a flat plane to rough in a round. Even a big, six-inch-wide cornice molding plane will work easily if you don't make it do the whole job or try to cut the entire profile on every stroke. Rock the plane to cut on one side and then the other. You may discover that you don't need to call the entire crew over to pull the tow rope for the finishing cut.

Roughing in the work with faster-cutting, easier-to-sharpen tools saves your planes, but eventually they will need sharpening as well. Some very excellent cabinetmakers sharpen their complex molding planes by honing just the flat side. Gradually, though, this makes hollows open up and thins the iron so it chatters. I don't think there is any shortcut. You simply have to get out the slipstones and sharpen on the bevel face of the contours and flats. Study the clearance bevels of each element of the profile and strive to maintain it. Bringing back one part means you have to bring back all the other parts as well. Each little curve and fillet in the plane iron is a cutter with its own needs, but it still has to work well with the others.

A sharp plane leaves behind a beautiful surface, but you can always help yourself by choosing better wood for moldings. Look for straight-grained stock with the growth rings perpendicular to the face. In any wood, the curves of the molding profile stiffen the shavings like corrugated roofing, so they shoot out straight from the plane until they buckle and fold of their own scant weight. When you finish planing, gather a handful of shavings and rub them hard up and down the length of the molding, burnishing it to a gleaming flow of curves and corners—"one intire Piece."

A final bit on planes. On September 26, 1810, Thomas Jefferson, sitting at his desk in the presidential mansion in Washington City, wrote a letter to James Dinsmore, an Irish joiner working on James Madison's house in the Virginia mountains. "Johnny Hemmings is just entering on a job of sash doors for [my]

The ogee plane works on the "spring."

In both furniture and buildings, the moldings come to life only under light and shadow.

house at Poplar Forest," he explained, "and tells me he cannot procede without his sash planes [that are] in your possession. If you could send them by Sunday's stage you would oblige me."

So here is Thomas Jefferson, asking for the return of wooden sash-making planes belonging to his slave, John Hemmings — an American who did not even own himself. In his will, Thomas Jefferson decreed that, upon his death, Hemmings would be free, so probably the last thing that John Hemmings made as an enslaved American woodworker was the coffin for Thomas Jefferson.

 Turner

*He was alone in his garret, imitating in wood one of those
indescribable ivory things, made up of crescents, spheres, inset
one in another, all lined up like an obelisk and just as useless. He
began the final part, the end was in reach! In the broken light of
the workshop, blonde dust flew from his tool like a fan of sparks
from the shoes of a galloping horse. The lathe wheels spun,
whirring. He smiled, chin lowered, nostrils opened, lost in the
complete, reliable happiness found in mediocre tasks which amuse
the intelligence by easy difficulties, appeasing it with fulfillment,
beyond which he did not dream.*

   —Gustave Flaubert, Madame Bovary, *1857*

Only a syphilitic French genius could dip into the flow state of woodturning and pull out the essence of bourgeois tedium. Woodturning was the once the hobby of kings. When Emperor Maximilian wasn't running the Holy Roman Empire, he was treadling away at his lathe. Tsar Peter the Great of Russia eagerly awaited every new attachment for his lathe made by his personal machinist. Through the eighteenth century, the woodturning bug spread among the upper crust of Europe. Aristocrats were turning pages as well, as great multivolume manuals of woodturning hit the shelves.

It didn't stop there. The bourgeois revolutions in Europe spread the woodturning hobby among the middling sorts. Readers understood Flaubert completely when he turned the hum of the hobby woodturner's lathe into poor Emma Bovary's water torture—another drip in the provincial ennui that finally drove her to suicide. If only Emma had taken up woodturning instead of opera and adultery! And if Flaubert had spent a little more time in the shop and less in the seamier boudoirs of Pigalle, he might have been a happier and healthier guy.

Too late for them, but our 2,000-year-old bargain with lathes still stands. For the great gift of perfect axial symmetry, all they ask is that we learn to work wood across the grain.

Let's first think flat. If you wanted to make a smooth hollow across the grain of a flat board, what tool would you choose—a scraper or a gouge? You'd choose the gouge, of course. The scraper could eventually do the hollow part of the job, but it sure wouldn't do the smooth. With the gouge, however, you could make shearing cuts across the grain, quickly shaping a smooth hollow—unless you chose the gouge but used it as a scraper, dragging it across the wood. Then you'd really have a mess!

But it happens all the time. When a length of wood is spinning in the lathe, it seems to take on an amorphous, plastic quality, but only our perception has changed—not the wood. Our persistence of vision makes the wood seem blurred, but the wood is persistent too—it's still cross-grained, and it wants to be treated that way.

So, when your friends pick up your turning gouge and stick it straight into the spinning wood, it may be the blurring that prompts them to use it as a scraper. Or, it may be fear. Even when you show them how to hold the gouge properly, it looks as if the tool is just on the edge of digging in and disaster. It is.

Time to stop turning and start teaching. Stop the lathe and turn the wood by hand as you show them how the gouge always cuts with the bevel rubbing. Show them that the gouge can never dig in as long as its rounded tip rides above the point of contact. Turn the wood by hand until they can make a cross-grain shaving curl unbroken from the skewed edge of the gouge. Then they're ready to start turning.

Centering

To ready the wood for the lathe, we first find the centers. Split the billet to keep the grain intact for strength and shave it close to size. Shaving a leg can

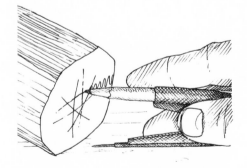

Find the centers with repeated strokes of the dividers.

mean roughly rounding it with a drawknife, or it might mean shaving it with planes to a square section for joining to the rails of a chair or table. In square-sectioned pieces, you can find the centers by drawing diagonals connecting the corners. You can find the center on any shape by repeated strokes with a marking gauge or dividers set at approximately half the thickness. Worked from all around the outside, the mesh of lines will bound the center.

Lathes also have centers. Treadle and great-wheel lathes (and others) have head stocks that support the drive center—a bladed spike that sticks into the wood to turn it. The other end of the wood spins on a polished, perfectly conical point—the dead center, mounted in the moveable tail stock. A spring-pole lathe usually drives the wood directly by a cord wrapped around it, so it uses two dead centers.

The lathe centers engage the centers of the wood, and a little preparation on the wood end helps them do that better. I always saw the slots for the drive center of the treadle lathe. On the dead center end, I either make a small depression with the point of the skew chisel or, in critical work, I drill a shallow hole with a small bit. A stroke of wax or grease to the dead center hole ensures easy and quiet turning.

The centers let the wood spin, but, at the same time, constrain it from wobbling or flying off. In metal lathes, you use a screw on the dead center (or ball-bearing live center) to adjust the amount of pinch given the wood. Wooden-bodied lathes may also have screws, but they work as well or better when treated like wooden planes. A tap at the base of the wooden tail stock moves its dead center in, and the wooden bed of the lathe flexes slightly, keeping the work piece tight yet free spinning.

Once in the lathe, there's a series of slow speed checks before the real turning begins. If the piece is curved in its length—centered on the ends but whipping around in the middle—you may have to reset at different centers to make a better compromise. The slow-speed turning also gives the centers a chance to set in. The tail stock usually needs a little more adjusting as you find the right amount of pinch.

Turning Gouge

As it is now, the piece may be just roughly rounded and out of balance. If turned at full speed, the wobbling and shaking would waste energy and could throw off the centering at the tail, enlarging it and making it run loose. You need to bring the whole length to a cylinder before doing anything else. This is turning, but not woodturning like you'll be doing once the piece is balanced.

A big gouge is the first tool applied to the wood. All turning gouges are sharpened with a flat bevel on the convex, outside edge, but their noses may be pointed or square across. Turning gouges for cutting coves need their rounded nose, but the roughing in gouge can be square across. This is often called a chair bodger's gouge after the fellows who work in the woods with their spring-pole lathes turning the legs for Windsor chairs.

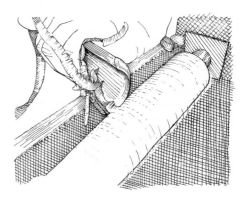

Round the stock with a large gouge.

Any gouge is useless without the tool rest, the moveable fulcrum that makes turning a precise art instead of a crazy battle. One hand holds both the shaft of the gouge and the tool rest at the same time. The other hand holds the end of the gouge handle. The tool rest puts the fulcrum point a fraction of an inch from the spinning wood. Big movements on the end of the gouge handle become precise refinements at the cutting end.

The first sweeps down the spinning wood with the gouge may not look precise, but they are. The gouge has to gradually take off the outermost, eccentric wood without diving in and catching. It's a matter of depth control with your hand acting as the fence, sliding along the tool rest as it firmly grasps the gouge. The other hand keeps the gouge tilted over and presented to the wood at an oblique angle.

Gradually these sweeps along the entire length allow you to set the tool rest closer, and you can begin bringing sections down to a proper cylinder. Starting three inches from the right end, you sweep the wood off to the end. As the wood becomes rounder, you gradually use less and less of the tight depth control as the gouge's flat bevel rubs against more and more of the wood.

Let's now venture down into the micro world where the action is. The edge of the gouge is shearing off a shaving and pushing it up out of the way. The shaving resists this and pushes back, tending to push the edge deeper into the wood. The deeper it pushes the edge into the wood, the thicker the shaving and the harder it pushes. The force cascades and the edge digs in.

Counteracting this force is the bevel, rubbing its broad, flat face against the face of the wood already cut. The wood, however, is not a flat surface, it's a cylinder. It only contacts the flat bevel in a line—because that's what cylinders and flat surfaces do. Still, that line is enough to support the edge against the downward push of the shaving.

But what if the bevel is not flat? What if the turning gouge or chisel has been ground and sharpened with a secondary bevel, or any other manner of a rounded bevel. Now we have a round bevel riding on a cylinder—the intersection is a point and not a line. Our foundation is now too small to resist the shaving's downward push. Forces cascade, the gouge digs in.

But your gouge, ground and sharpened with a flat bevel, shears off a shaving that spirals up the channel of the gouge, flowing out over your hands onto the ground. When each three-inch section is done, you move along and round the next section until the whole piece is a smooth cylinder.

Skew Chisel

Sometimes, of course, you may not want an entirely smooth cylinder, such as when you intend to leave square pummels in a piece for mortise and tenon joints. In such cases, you start with a square-sectioned piece and turn only some of it to a smooth cylinder. To protect the decorative turnings from damage, the mortises should already be cut in the pummels before the piece goes in the lathe. Once in the lathe, it's the pummels that need protecting. Their sharp corners can easily splinter off unless you first isolate the square sections from the round with a cross-grained knife cut. It's the same cut that you always make. Here, instead of the striking knife, the nicker on the fillister plane, or the slicing teeth on the big crosscut saw, you use the skew chisel.

The skew chisel has two flat bevels with the edge shaped obliquely to its length, giving it a long corner and a short corner, or a point and a heel. Set the skew chisel on the tool rest, held vertically on edge with the point down. Because the wood turns toward the tool from the top, this creates the same relationship that you would give a knife when drawing it across the grain to score a line. Square up to the division line between square and round and start the wood spinning. Lightly make contact and then stop to see that you are cutting in the right place. You can't cut very deep by striking straight in at the same place, any more than you can with an axe on a tree. When your first cut is still just nicks in the corners, cut back into it from one side or the other, or both, widening and deepening the cut each time until you reach the depth of the circle within the square. If you want a slightly rounded transition from the pummel to the round, now is the time to create it. Work with repeated fine cuts using just the point, always tilting slightly away from the surface with the rest of the blade.

Beads and Coves

With the work cylindrical and any square sections isolated, you can begin the pattern work. Like any molding, turning uses beads, coves, ogees, ovolos, and such. Copy work is a good way to learn, but see that you are copying a bold original turning, not something that is already a vague copy of a copy. Spindle turning is everywhere, and if you have never turned a critical eye to legs, it's time to begin.

If you are working from a pattern or a finished piece, a pencil, a set of calipers, and a parting tool can be your guides. Choose transitional spots, the bottoms of coves, the tops of ovolos, the ends of a set of beads, and mark them on the turning stock with a pencil. Set the calipers to the diameter of a hollow on the pattern and strike in with the parting tool until the calipers tell you to stop.

The parting tool is, again, a chisel, not a scraper. Think about the cross-grain and present the point so the edge cuts with the bevel rubbing. This puts the tool riding on the face of the wood as it turns toward you, a shaving flowing from the top. As the diameter of the wood decreases, adjust the angle of the parting tool to keep it cutting. The calipers receive a thorough high-speed shaking each time you test the diameter of the revolving wood. This vibration can cause the adjusting screw on the calipers to back off, enlarging the setting. Re-check your calipers against the model before you move on to the next sizing point.

The cut of the parting tool is square to the cross-grain and rough. But it's just the depth guide—now come the shearing cuts with the gouges and skew chisels. As always, they cut with their bevel rubbing, and they always cut downhill, always from the larger diameter down to the smaller.

Say you want to cut a cove. Start at the left side and turn the gouge so that the bevel faces left and the nose rides clear of the wood. The edge then meets the wood obliquely and can shear it into a corkscrew shaving. Arc down into

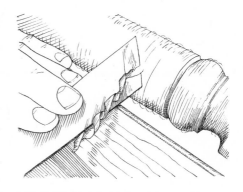

Smooth the cylinder with the skew chisel.

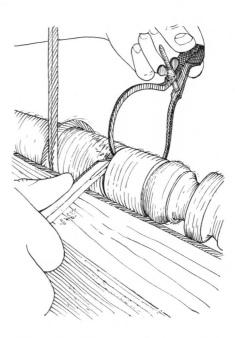

Strike in critical diameters with a parting tool and a pair of calipers.

Work into a cove by cutting down the slopes of the sides.

the cove with a shallow shearing cut to the middle. Now start at the right side of the cove and meet the first cut at its bottom. Repeat the cuts from left and right until you have the shape you want.

It's the same all over. The neck of a vase shape is a gentle oval cove. A tight cove may benefit from defining the edges with knife cuts from the skew chisel, but the gouge work is the same—alternating cuts from either side, always from the larger diameter to the smaller, the point of the gouge riding clear.

There are convex shapes in turning as well, but they're only created by cutting wood away. Beads in turning are like beads in joinery—half circles rising from the background. Here, the beads are running across the grain, so strike in their margins with the point of the skew chisel as you work. The heel of the skew does the work, starting flat on the middle of the bead and then rolling around and down to shear one side round. Flip the skew chisel over, angle it the other way and roll over the other curve. Small beads take two moves, larger ones more. A gouge can do a bead as well, rolled over and swung around in the same manner, but the skew is the smoother tool.

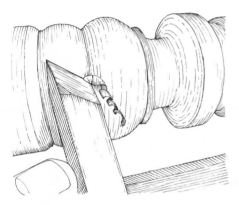

Turn a bead with the heel of the skew—always working from the larger diameter to the smaller.

Balls are another matter. Those who regularly turn spheres use special holders that spin the block at changing angles as the sphere emerges. If you have just a few to do, say for ball and socket joints, start by turning a cylinder to the same diameter as the ball. Mark out the width, which is always the same as the diameter. Strike in on the waste sides of these lines with the parting tool, leaving enough wood to keep the piece turning steady. Find the center and mark it with a pencil. Turn each hemisphere with the gouge at first and then refine with the skew chisel. From the Equator to the Tropic of Cancer, work on top of the ball with the heel of the skew. From the Tropic of Cancer to the North Pole, work from under the ball with the point. Repeat on the Southern Hemisphere and part the last bit of wood with the point of the skew.

Finishing any spindle turning also relies on the skew. Work only with the lower half, the half toward the heel, keeping the point riding high and clear. All but the tightest hollows receive a down-the-slope refining pass of the blade. Finally, a handful of shavings pressed against the work burnishes it to a gloss, leaving the edges distinct rather than rounded and blurred.

Offset Turning

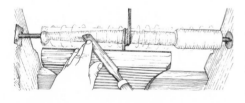

Turn the pad foot first, then offset the leg and turn to the "ghost."

Sometimes, though, the work isn't as round as it might be. If the work gets thin enough to bend away from the cutting tools, you'll have to back it up somehow. There are various steady rests to mount in the lathe, but like most turners, I rarely use one. When the work needs backing up, it's far easier to reach around with your fingers and hold it while your thumb keeps the turning gouge or chisel pressed to the tool rest. I'm sure there is a way to injure yourself doing this, but using foot-powered lathes, I have yet to find it.

Another instance of intentional nonround turning is found in furniture legs, which can be turned with offset pads, or, rather, offset legs, by turning the piece twice. The first turning shapes the leg as a full cylinder with the pad centered on the end. Then, the leg gets new centers, offset as you wish the leg to be offset. On a rectangular table, you probably want the leg to recede from both corners and then come back out at the bottom to sit plumb below the corner. This requires you to offset the new center at the pad end diagonally, halfway to the inside corner. On the top end of the leg, offset the new center just a little in the opposite direction. This offset brings the end of the square pummel and the start of the round leg back to the center of rotation.

Looking at the spinning piece, you can see through the ghosts to the tapered solid leg at their center. Turning the offset leg follows the same guidelines as roughing in—most of the time you're cutting air. Work down through the ghosts to the leg, leaving the pad intact. The pad needs some final work, making for a third turning, this time back at the original centers. Rounding the bottom of the pad removes the wood used for the offset center, so it comes last.

I called this nonround turning, but in truth, at any point in the cross section of these legs the wood is perfectly round. Oval turned tool handles are truly oval, or close to it. Actually, it's the wood left by three turnings at different centers. The first round turning defines the long axis of the oval. The subsequent offset turnings remove the wood from the sides to shape the short axis of the oval. The remaining wood is the intersection of three circles and needs just a little smoothing with a cabinet scraper.

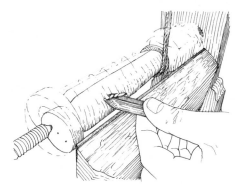

Two offset centers will shape an oval handle.

Ball and Socket

We spoke of balls before. Usually these are decorative, but there is a style of embroidery stand that uses a ball and socket head tightened by a wooden hand screw. The ball takes some skill to cut, but the socket in the end grain of the pedestal can be shaped by engineering.

First, however, consider how you would do it freehand. When you want to bore the length of a spindle, say to make a flute, you finish the cylinder, then remove the dead center and replace it with a hollow center. This can be a sharp ring or a funnel-shaped socket where the newly rounded end of the cylinder can ride. Push a long auger at the exposed end grain in the center of the hollow center and you can bore down the length as the wood spins. You could shape a socket the same way, spinning the end of the pedestal in the hollow center as you reach in with a gouge and scraper to hollow it out like a pumpkin.

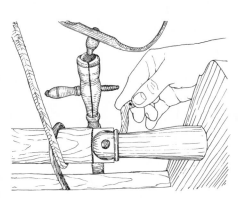

Cut-away view of the scraper hollowing the socket for a ball-and-socket embroidery stand.

The last scraper to make the socket could be a side-cutting hemisphere, ground exactly to the final size. That is what I use, but instead of working freehand, I have the scraper mounted within a wooden nose. The nose just fits into an auger hole bored into the end grain of the pedestal. The scraper is double faced—a disc with two faces ground off. It fits entirely within the nose in one orientation, but when rotated, the rounded faces are gradually exposed, scraping their profile in the cylindrical walls of the auger hole. When the socket is fully formed, you rotate the scraper back to fit within the nose, take off the pedestal and shake out the dust.

This gives you the socket, but the ball is 1/4 inch larger than the opening. To get the ball in, saw a slot right down the middle of the pedestal. This lets

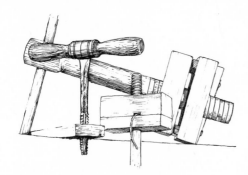

You can cut wooden screws with taps and dies...

...or "chase" them on the lathe.

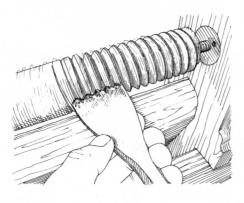

Lacking a faceplate, you can turn tabletops and bases while they are mounted on their shaft.

the sides of the socket spread as you press in the ball. The ball is in, but it's loose—until you make the wooden screw to turn it tight.

Wooden Screws

Cutting wooden threads is always associated with woodturning. At the very least, the screws begin as cylinders before threading with hand screw boxes. Eighteenth-century woodturning manuals give thorough instructions on making screw boxes and other devices for cutting both large and small screws. I have included translations of these in the Appendix as Plans A and B.

Along with the screw boxes and special screw-cutting lathes, these books illustrate chasing tools—scrapers with teeth spaced at the pitch of the desired thread. As the wood turns, you deftly sweep the tool across the cylinder and strike a set of spiraling threads. Once the first threads are cut, they guide the chasing tool along, cutting subsequent threads down the cylinder.

Chasing threads is more commonly done in smaller works such as the internal and external threads for lidded boxes, but you can also chase the threads for a two- or three-inch bench vise screw. Diderot shows such a large chase in the section of the *Encyclopédie* on turning, and an excellent blacksmith forged a duplicate for me using the scale provided in that eighteenth-century book. The five scraper teeth are 7/16-inch equilateral triangles, all sharpened as scrapers except the rightmost one. I dubbed over that tooth's edges to keep it from cutting, but retained its shape. This safe tooth can ride against an existing thread and push the cutting teeth along, advancing the spiral.

The five scraping cutters create beaucoup drag, but the spring-pole lathe has the power to roll a dried hickory blank right through it, sliding the chase down the tool rest with every turn. In theory, you can strike the first thread with that single deft sweep of the tool. I don't even try, and give the first few inches of the screw a hand-sawn and -rasped starter thread—as explained in the translation of Salivet's description of cutting large screws. Once the chase gets a purchase in the first spiral, every turn pushes the chase to the left, duplicating the first thread. Held lightly at an angle, the teeth to the left scratch a gradually deepening spiral. As each section reaches full depth, you give the chase new footing and copy that completed section farther down the screw. The threads cut by the chase are only as good as the first spiral. If there's a wobble in it, that wobble is copied all the way down the length of the screw. The old masters called this a "drunken screw."

Bowl Turning

So, it seems, there are times when you need a scraper. Sets of turning tools include scrapers, and there are times in spindle turning when you're glad to have them. Perhaps when turning a boxwood flute you need a sharp corner. The scraper works brilliantly on box and other very hard woods—if it's sharp and held at the proper angle to the surface.

A spike mandrel lets you turn bowls between centers.

It's the burr on the edge that does the cutting, and some scrapers take a decent one just from the grinding. The bevel angle of a scraper is about 60 degrees, and the top face is flat. You can also sharpen the scraper with a turned burr after honing smooth the bevel and the top. Set the scraper upright in the vise and pull a rounded steel burnisher, perhaps the back of a gouge, across it. Scrapers made from old files need considerable force to turn the edge.

I have gone on about never using a chisel or gouge as a scraper, but the inverse is more dangerous. Scrapers are intended to drag lightly across the surface—the handle end up and the cutting end down. Pointing a scraper up into the wood can make it catch and go flying. Worse, it could snap the brittle scraper and put your eye out. Keep your scrapers sharp and work with a light touch.

Scrapers also finish the job in plate and bowl turning. I do not have face plates on my lathes, but most items that call for a turned plate also include a

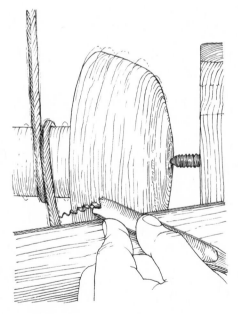

Start in the cut in the temperate zone and work toward the Equator or toward the pole.

shaft. These are things like small tables, candlestands, and such. In these cases, it's easy to mount the disc on the shaft and turn it between centers. For bowls, use a spike mandrel driven into the face of a hewn blank. The mandrel gives room to work the bowl between centers, and in the case of the spring-pole lathe, gives the drive cord something to grasp.

Bowl turning with scrapers would take forever—so gouges do most of the work. In bowl work, you're cutting across the grain, against the grain, and with the grain on every turn. On the convex face, work with the bevel of the gouge rubbing as you work from the bottom to the rim. The long axis of the gouge always rides close to a tangent of the surface that you're cutting. On the inside face, you're cutting the inside walls of a doughnut shape. Again, the bevel rides the tangent of the curve. Turn the bowl by hand in the lathe and find the angle where the gouge shears the wood without digging in. You can turn with that.

The central core stays in the bowl after the first turning, because the wood is still green at this point. Now it has to dry, shrink, and harden up before it can go back in the lathe for the final turning, including a finish with the scrapers. It's all up to you, though. You can turn bowls at one go in fresh-cut wood and let the distortion from the uneven shrinkage be part of the charm. Or you can just roughly round the bowl blank and set it aside to dry before any real shaping begins. The green wood is heavier, but easier to cut, and easier on your edges.

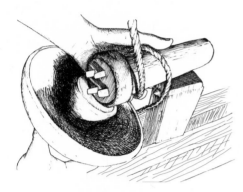

Remove the spike mandrel and knock out the core left in the bowl.

Gouge Grinding

Whatever wood you cut, eventually you'll need to sharpen your tools. The bevel on the bowl turning gouge is shorter than for spindle turning, and the nose is usually square across. Otherwise they are sharpened in the same way. Grind the bevel on a sandstone wheel or a tool grinder that lets you work on the flat side of the stone. On a spindle turning chisel with a longer nose, you need to swing the back end of the handle around as you rotate the gouge on its long axis. I want to tell you a specific angle for grinding, but a wide range works. Looking at the bevels on my chisels and gouges, they seem to run from 30 to 50 degrees. Each different bevel requires that I hold the tool at a different inclination to get the flat of the bevel rubbing, but they all cut.

When you hone a gouge or turning chisel, the bevel must remain flat as it was ground. Hold the tool in one hand, resting at an angle on the edge of the bench or tool rest. Rotate the tool slowly as you work the whetstone back and forth flat across its bevel. The whetstone can slide directly on the bench if it's mounted in a box. You can just as well hold the stone in your fingers and slide the back of your hand back and forth across the bench. Just keep the bevel flat.

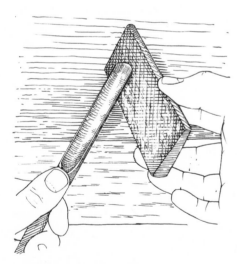

Hold the gouge against the corner of the bench top and slide the stone back and forth.

On the inside of the gouge, again the stone moves as the gouge turns. Hold the gouge near the point with your fingertips, the handle hanging. Hold the round-edged slipstone lightly in your other fingertips as you work it up and down in the channel of the gouge. The light grip lets the stone ride flat as you slowly roll the gouge to ensure that the whole hollow gets honed.

Pulling the cord turns the bobbin, twisting the bow strings and flexing the bow.

Spring-Pole Lathe

There are differences in the way lathes are driven, their potential speed and power, and the way the wood fits in them, but the principles of successful turning remain the same with any lathe.

The spring-pole lathe works with reciprocating circular motion—the foot treadle makes the cutting stroke and the spring pole handles the return. Between the pole and the treadle, the drive cord wraps around the work, passing on the side facing the turner. In a handy bit of mechanical serendipity, the spring-pole lathe matches its drive torque to the size of the work piece. The larger the piece that you're turning, the larger the diameter where the cord wraps around. This "bigger pulley" costs speed, but it gives you more leverage. On smaller diameter pieces, when you want speed and need less torque, the cord wraps around a "smaller pulley," and the work flies. If the circumference of the work piece is six inches around, then every foot of treadle travel on the cord end gives you two full revolutions. If the circumference is half that, you get four revolutions and twice the speed.

A spring-pole lathe works fast, but not as fast as a wheel-driven lathe. It does, however, have slow speed torque that flywheel lathes can't match without a mess of reduction pulleys. The treadle also gives you great control. You can kick it out for more speed and less power—or set your foot closer to the drive cord end for more torque.

Of course the wood has to go backward as well. You cut on the down stroke with a reciprocating lathe and let the spring do the return. I suppose you need to make some movement to clear the tool on the return stroke, but it's not anything that you think about. It's a tiny movement—if any—that becomes natural after your first five minutes.

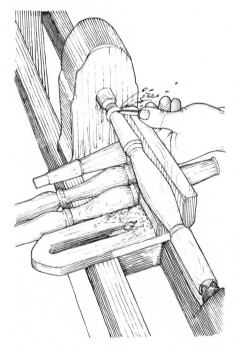

The pole of a spring-pole lathe can be just that, but use a dry, springy pole and not a green, limp one. A bow with a roller strung on a doubled bow string makes a more compact drive. The drive cord wraps around the roller, so that as the down stroke of the treadle pulls it, it turns the roller and twists the doubled bowstring like a Spanish windlass, bending the limbs of the bow. I have taken to using a lever-action lathe with two poles mounted underneath the bed. It is compact and portable. I love it, and you can learn how to make one by reading Plan E in the Appendix.

Treadle Lathe

Leonardo da Vinci's sketchbook contains perhaps the earliest depiction of a continuous-action treadle lathe. The essential element of the treadle lathe is the crank mechanism that converts the reciprocating action of the foot treadle into the rotary motion of the large flywheel pulley, which, in turn, drives the smaller pulley in the head stock at high speed.

Even this simple mechanism is prone to disorder. The weakest point in my design is the connection between the 1/2-inch-diameter axle and the large driving wheel. To assemble the lathe, the axle/crank has to be removed from the flywheel. It must then connect to the flywheel with enough strength to resist the powerful shearing forces between the little axle and the big wheel. I use a shear pin through the axle and the flange mounted on the drive wheel, and when a new user gets out of phase on the treadle, the shear pin shears.

Still, after 30 years, I've had to make a new axle only once. If you want to make one like it for yourself, see Plan F in the Appendix.

9 Cabinetmaker

*[Cabinetmaking is] joinery of a superior description,
working with finer tools on more costly woods, and
producing more sightly effects.*
—Mechanics' Own Book, *1885*

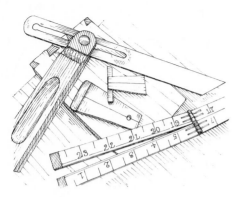

Dovetail markers and a sliding bevel.

Cabinetmakers seem like such nice people, but who cuts the walnut's crotch and exposes its storm-stressed flames in a window of polished shellac? Who stuffs balls into the claws of beasts and makes them into a chair leg? Even the gnarliest burl is helpless as they saw it into slices, level it with toothed scrapers, and choreograph the pieces into a symmetrical dance. Cabinetmakers capture wild natural beauty and cage it within classic geometry—yet they hide their own sweat within secret joints that show nothing of the inner connections.

In Georgian cabinetmaking, the sightly effect was to reveal the wood but conceal the joints. Contemporary cabinetmaking comfortably reveals both. Dovetails give you that choice. They can be fully exposed, half-hidden, or kept secret. Dovetails are fundamental to cabinetmaking, but wood is wood, and the mechanics of joining it spans all the trades. Now, the multiple-dovetail joint has transcended even its functional identity to become the icon of handcraft in wood.

But the mechanics never go away. When you pull out a drawer, you see splayed dovetails on the sides and trapezoidal pins on the end grain of the drawer front. You certainly could make a dovetailed drawer with the tails on the front and the pins on the sides, but such a construct would orient the weakness of the joint in the same line as the greatest load. Dovetails are the strongest of joints, except in one direction.

So, first, consider mechanics, even if you let aesthetics make the final decision. Second, how much joint do you want exposed? A Charleston desk joined with secret miter dovetails shows no joint at all. A Shaker desk with through dovetails shows joints at all corners. In their drawers, however, they all look the same, both pieces joined with half-blind dovetails. We'll take each of these joints in turn, starting with a through dovetail for a chest.

Through Dovetail—Tails First

Planing the ends of the tails on the through dovetail joint.

Dovetails are cut by superimposition. I wrote earlier about laying out mortise and tenon joints by superimposing one element on the other and transferring dimensions. This makes some people uncomfortable. Good cogs in the mass production machine are supposed to work from a measured drawing, manufacturing each precise part and then fitting them all together. But try to find a measured drawing from the days of the great cabinetmakers—there are none. Their forms emerged from the constraints of classical proportion, customer demand, and mechanical necessity. They built furniture largely with superimposition shaping each piece of the emerging whole.

The immediate question for dovetails is, which do you cut first? Which part gets to be the super of the superimposition, the tails or the pins? In secret miter dovetails, we have to cut the pins first, so that's settled. But we're making through dovetails now, so we'll do them tails-first.

The wide boards of this chest will have lifting handles on the end. Thus, the tails will go on the sides and the ends will get the pins. Dovetailing can be no more precise than the boards are square and true. Square all the end grain with a block plane, leaving it smooth to allow clear marking.

Assuming you are joining equally thick boards, set the gauge 1/32 inch greater than the thickness of the boards and run the fence of the gauge against the end grain of each of the boards to mark the extent of the tails and pins. This extra 1/32 inch gives you something to plane off the completed joint. Lightly trail the point of the gauge across the grain rather than scratching and tearing. A lightly held cutting gauge serves even better.

Now decide the spacing of the tails and pins. If the chest has any skirting around the base or lips to meet the lid, consider their locations so the dovetails won't begin or end awkwardly. If the boards will have a groove plowed down their inside to hold a bottom, that groove should not overlap the joint between a tail and pin. Make these grooves first so they won't be a surprise later on.

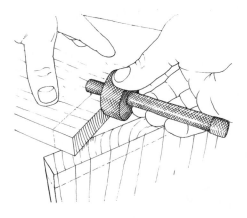

Gauge each piece to its mate's thickness, plus 1/32 inch.

To fight confusion, I'll call the piece with the tails the tail board, and the one with the pins the pin board.

We will cut the tails first, meaning we are working on the broad sides of the chest and defining the sockets for the pins between the tails, the pin spaces. Custom demands that dovetails end on a half-pin, although you may find otherwise in old pieces.

The size of the pins relative to the tails is an aesthetic/mechanical tradeoff. For a heavy chest, you could make them equal in size, but even when equal spacing would be stronger, a chest looks far better when the pins are about half as wide as the boards are thick, and the tails are about three times as wide as the pins.

By this guideline, the thickness of the wood determines how many tails and pins will fit within a given width. For 3/4-inch-thick, 11-inch-wide boards, this comes to 8 dovetails separated by 7 pins, bounded by a half-pin on each end. We thus need to divide the 11-inch board into 8 equal spaces—not measuring from edge to edge, but from the centerline of one outer half-pin to the centerline of the other.

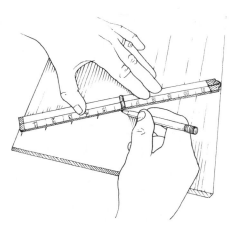

Divide the width into equal parts using a diagonal ruler.

At their widest points, these half-pins should be the same width as the full pins, at least half the thickness of the wood, in this case, 3/8 inch. This puts their centerlines at 3/16 inch back from the edges. (If you think the corners are going to take a beating, you can make the bounding half-pins a bit wider.)

Back to dividing an 11-inch board into 8 equal spaces. Hold a ruler diagonally on the board with the zero end crossing the centerline of one half-pin and the 12 crossing the centerline of the other. Put a pencil mark every 1 1/2 inches down the diagonal ruler. Carry these marks up to the end using a try square and a pencil, dividing the width into 8 equal parts.

This works for any width or number of divisions. You would place a mark every 2 inches if you wanted to divide the board into 6 equal parts. For 7 equal parts, position the ruler at zero and 14 and mark every 2 inches. For 5, go to 15 and mark every 3.

Anyway, where each pencil line crosses the line marked by the gauge, measure out the width of the pin space, 3/16 inch on either side for a total of 3/8 inch. This is the widest part of the pin space, the narrowest part of the dovetail, and the slope starts here.

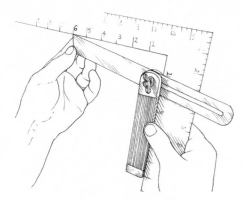

Set the sliding bevel for an angle of 1:6.

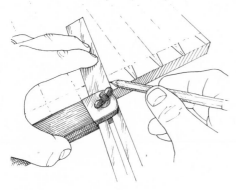

Lay out the spaces between the dovetails.

Saw the sides of the waste spaces.

L E F T : *Chisel down from one side . . .*

R I G H T : *. . . and complete from the other.*

For most work, find the slope angle for your dovetails by setting the sliding bevel to cross the one and the six inches on the square. A one-in-five slope looks more robust; a one-in-seven looks more delicate.

Whatever angle you decide upon, set the beam of the sliding bevel on the end grain and draw the converging sides of the pin spaces up from the gauged line with a sharp pencil. Mark the little spaces with Xs so you will know what you will be cutting away. Carry all these lines square across the end grain and use the bevel again on the back side. The lines across the end grain are the most important. You can have variation in the slopes between different dovetails—they'll just look funky. But if the lines aren't sawn square across the end grain, the assembled joint will have gaps that weaken the whole works.

Sawing at last. A dovetail saw is simply a fine-toothed backsaw, but use what you have. Saw precisely parallel to the lines, touching them but leaving them intact. Saw diagonally at first and watch that you leave the line on both faces.

The ends where the half-pins fit will come off with two more cuts from the saw, but the rest of the pin spaces need chisel work. You could remove most of the wood with a coping saw or, in heavy work, use an auger, but normally this is all push chisel work. If the pin spaces are 3/8 inch wide at their base and taper down from that, the largest chisel you can use is one about 5/16 inch wide and beveled on the sides because you must start a bit clear of the scribed line.

Try working with two chisels, starting by pushing straight down with the wider chisel, flat face to the line. Then, with the narrower chisel held bevel down, push toward the bottom of the first cut and remove a chip. Working down, excavate halfway through. Flip the board over and make the same V-opening to meet the first one. Leave the narrow end intact to support the waste until the V-cuts meet—otherwise it might break out and tear the wood. Clean up and trim the pin sockets back to the line.

Now we'll cut the pins to fit into these sockets.

See that the board for the pins is not cupped, clamping a batten onto it if necessary to hold it flat. Set it in the vise as you hold the end grain flush with

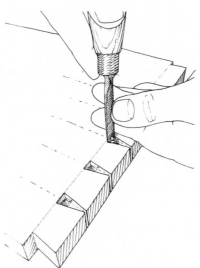

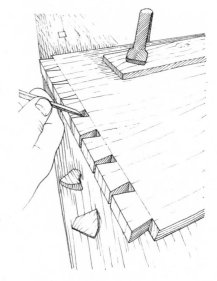

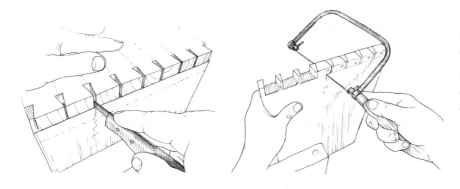

LEFT: *Set the tails on the joining board and mark the pins.*

RIGHT: *Cut the sides of the pins with a dovetail saw, and remove the waste with a coping saw.*

a piece of scrap or a plane lying on its side. Push this scrap or plane back across the bench to support the back end of the tail board, the super board of the upcoming superimposition.

Align the board bearing the tails on the end of the pin board. See that you are well and truly aligned to the scribed line with the extra 1/32 inch hanging over. A pencil can't reach into the tiny pin spaces, so scribe the width of the openings onto the end grain of the pin board with a knife, cutting dead flush to the walls of the sockets. Don't let the board move until they are all marked.

Remove the tail board and elevate the pin board enough that you can carry the lines square down to the depth mark with the try square. Mark the waste pieces with penciled Xs. You are now about to remove big pieces and leave small ones. Accurate sawing on the waste sides of these lines will make the joint fit without trimming. Split the lines with the edge of your saw kerf, and the whole should fit up fine.

The waste from the broad sockets for the dovetails is large enough to make the rough cut with a coping saw worthwhile. Work from the face side with the teeth cutting on the pull stroke. It may be that your coping saw cuts a wider kerf than the dovetail saw, so keep it pulled to the waste side as you work your way down to make the turn across the grain.

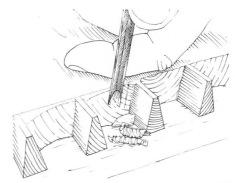

Chisel flush to the gauged lines.

The broad bottoms of these sockets for the tails form visible parts of the joint and need to be straight and clean. Lay the piece on a scrap block on the bench and pare toward the middle with the freshly sharpened chisel set right in the depth line for the final cut. Finally, you stand the piece up in the vise again and shave with a diagonal shearing cut of the chisel to smooth the bottom.

Test the fit. Don't bevel the ends of the pins to make the fit easier — you've got only 1/32 inch to play with. Instead, pare back the hidden, inside corners of the dovetails. This beveling strengthens the edges of the tails and gives a little clearance for any stuff in the corners. Watch for a fat fit that could cause a split, particularly against the half-pins at the ends. Use a cabinet file to trim if you need to, but mind its corners.

Set a batten across all the tails to spread the impact and protect the tails as you drive them up with the mallet. You may want to use bar clamps to draw the joint up instead. First time out, bring the joint all the way closed before tapping it apart, applying glue and re-closing and squaring it. With more experience you can just make a partial test fit before gluing. After the chest is joined and

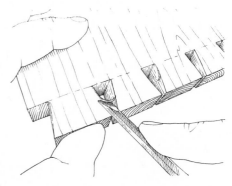

Bevel the inside corners of the dovetails.

Tap the joint closed.

"Kerfing in" a box made with mitered corner dovetails.

the glue fully set, trim the protruding ends of the tails and pins with a block plane, working from the outside in so you don't splinter off any ends.

Through dovetails in plain flat boards are the simplest expression of the joint. In practice, there are complications.

If you have a groove on the inside of the chest, you must leave an extra shoulder at the bottom of the dovetail socket to fill the wood removed by that groove. This makes a shallower, slightly weaker and odd-looking joint. You may be better off stopping your grooves short of the ends of the chest so the dovetails can remain whole. If you have a rabbet running around the inside of the rim of the chest, leave an extension of the dovetailed piece to fill the gap in the half-pin.

You can also deal with a rabbet by cutting a plain, butted miter into the half-pin of the dovetailed joint. This miter allows you to cut uninterrupted exterior moldings as well. Your brain has to change gears to do this, from square cuts to diagonal. Cut the miter a little fat and then kerf it in by repeatedly sawing the crack and tapping it together until the dovetails close.

Lapped Half-Blind Dovetail

These are the dovetails for drawers and casework, hidden on one side of the joint and exposed on the other. On the carcase of a desk, the top and bottom may be the same thickness as the sides, but their dovetails stop short, leaving unbroken the display of grain on the sides. Drawers also have an unbroken display on the front, but usually join pieces of unequal thickness—the sides and back being thinner than the front.

Drawers often have their bottoms set in grooves plowed into the front and sides. The bottom fits in this groove like a door panel, fastened only at the front and left free to expand to the back. The back of the drawer is cut short, thus allowing the bottom to move freely beneath it. At least that's one way (some use little battens), but if you do plan to set your bottom in a groove, plow it into the sides and front before laying out the dovetails.

Lap dovetails show only on the sides of a drawer.

Only the front of a drawer has half-blind dovetails. The back connects with common dovetails, and since we just covered that, I'll mention the back only as it arises in the sequence of drawer layout and work.

Lapped half-blind dovetails for drawers join the thinner sides to the thicker front. Because the sides are usually thinner, and because they take the tails, you can clamp several drawer sides together and saw them at the same time. That makes this tails-first dovetailing, as before, but with a different method of transferring the tails to the pins.

Set your marking gauge to the thickness of the drawer sides and back. With the fence of the gauge always riding on the end grain, run it around the ends of the drawer back, around the ends of the drawer sides that will join to the back, and then just on the interior side of the drawer front.

Reset the gauge for about four-fifths of the thickness of the front and, fencing from the interior face, mark it along the end grain of the drawer front. Finally, run this setting around the ends of the sides that will join to the front.

Old drawers may have just one or two big dovetails and two or three small pins. The regional variations are worth your study as well, so whenever you visit a fine home with locally made furnishings, always ask your hosts if you may examine their drawers. There is great individual variation within the larger historical trends.

Free from historical precedents, your own drawers may have any number of dovetails, and you lay them out in the same way as we did earlier. Include the groove within the body of the lower dovetail and it will not show in the finished drawer. Make the widest point of the pin sockets equal to half the thickness of the drawer front. A one-in-six angle is always appropriate.

Before you square the lines across the end grain, lay out the dovetails for the back of the drawer. Mark all the waste wood with Xs. Then, clamp the two drawer sides together and carry the lines square across both pieces. Drawer sides are rather thin, and if the other drawers are the same size, let them join the pack. You can gang saw dovetails—but not pins.

Make all the sloping saw cuts on the drawer sides with a fine saw, but don't take out the waste wood just yet. You need the saw kerfs intact to transfer the dimensions to the drawer front.

Mount the drawer front upright in the vise with the groove away from you and superimpose the appropriate drawer side upon its end grain. Align the end of the drawer side with the line scribed along the end grain of the drawer front. Hold the side firmly in place as you place the saw back in each kerf in turn and draw it lightly back, leaving lines, transferring the dimensions to the end grain of the front.

Turn the drawer front around or lay it on the bench top and use a try square to drop perpendicular guidelines down from the edge. Mark the waste with Xs.

Place the drawer front in the vise with the inside toward you for sawing. You can't saw the entire cheeks of the pins, but you can saw the diagonal. Your diagonal saw cuts may even extend into the visible parts of the interior of the drawer, but never through to the front.

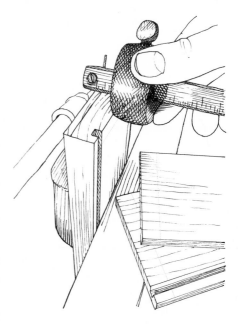

Gauge the lap on the end grain of the drawer front and around the ends of the drawer sides.

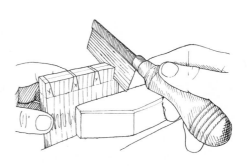

Lay out the tails on the sides and saw them clamped together in pairs.

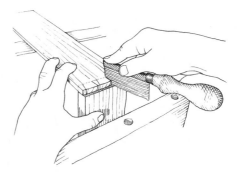

Align a side on the front and draw back the saw in the kerfs.

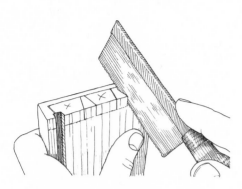

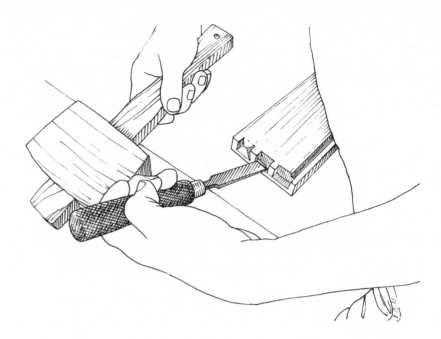

ABOVE: *Saw as much of the sides of the pins as you can.*

RIGHT: *Sit on the drawer front and alternately chop down and split out the waste with a chisel.*

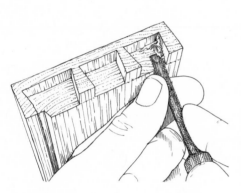

Finish the spaces between the pins with a paring chisel.

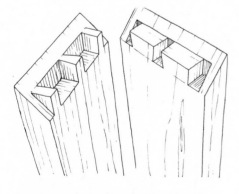

Secret miter dovetails are hidden when the joint is assembled.

As always, you want to saw just on the waste side of the lines, but here it's perhaps a little harder. Drawing back the saw in the kerfs left accurate marks, but they are shallow grooves — very attractive to your saw. You must start and continue the saw on the waste side of these grooves, not in them.

The rest comes out with a chisel. Lay the drawer front flat on the bench and either pare back or chop out the waste with your chisel. Drive in cross-grain cuts near the line and then split back to these cuts from the end grain and repeat until you reach the full depth. A skew chisel is handy to reach into the angle of the corners. For the final trimming, you may again find it more convenient to clamp the work upright in the vise and make bevel-up slicing cuts across the grain.

Finish off the drawer sides and back as we did earlier on the through dovetails. As your final cut before assembly, slightly bevel the inner corners of the tails to help them ease into their sockets.

Slip the bottom into place to hold the drawer square as the glue dries. You can glue the bottom into the groove in the drawer front, but never along the sides or back. Of course the long grain of the bottom board runs from side to side, directing any expansion from front to back.

Secret Miter Dovetail

Before you close the joint, secret miter dovetails look like cliff dwellings built on the slopes of the miters. However, when you bring the pieces together and bunt them with your hand, they close with a clack and the cliff dwellings vanish. There's no sign of any joint, no trace of the work inside — just the thin line of the miter.

The work begins with the pins. The form of secret miter dovetails dictates that you cut the pins first and then transfer their dimensions to the tail board.

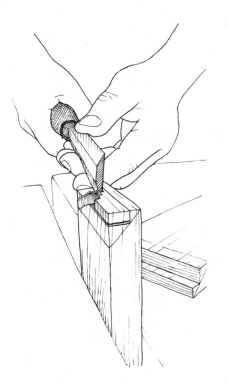

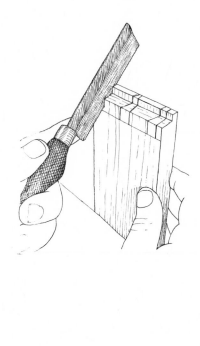

LEFT: *Saw and chisel equal shoulders.*

RIGHT: *Lay out the pins and start the sides with diagonal saw cuts.*

Gauge the inner faces of both pieces to the thickness of the wood, with no allowance for trimming off. On the edges, connect the outside corner and this gauged line to mark the miter line. Use a knife for this line to protect it from splintering. It's a line that will show.

Now gauge the rabbet that defines the shortened pins and tails that hide in the miter. Set the gauge for one-fifth of the thickness of the wood. Fence the gauge against the outside faces, scribing across the end grain and around the corners to touch the miter line. Now set the fence on the end grain and carry the line around the edges and the inside faces.

Saw out the rabbet on the waste sides of these lines. True it with a low-angle shoulder plane, taking care not to go over the miter lines cut on the edges. The shoulder plane is an expensive rabbet plane for end grain. It gets its name from lying flat on the cheek of a tenon and planing the end grain of the shoulder.

Lay out the pins on the end grain of one piece. You can't use the sliding bevel or an ordinary dovetail template to lay out the angles of the pins on the end grain—the shoulder gets in the way. You can snip a guide out of tin, but if you're ready to cut secret miter dovetails, you're also probably able to cut a good dovetail angle by eye.

As with the half-blind dovetails on drawers, you can start the pins by sawing some of their cheeks on the diagonal. Remove the waste with the chisel.

Shear the small shoulders at either end of the joint to the miter, but leave the long square lip until after you have transferred the pin spaces to the mating piece.

Set the end grain of the pin board on the tail board, using the lip and the scribed line as guides. Reach in between the pins and scribe with a knife to delineate the tails. Saw the diagonals and chisel out the waste.

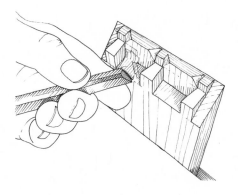

Complete the spaces between the pins with a chisel.

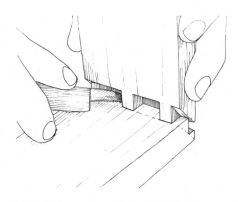

Align the pins on the mating piece and scribe around them.

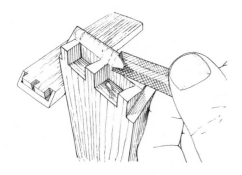

Complete the tails and pins, then shear the miters.

Check the fit of the joint. It will come four-fifths closed before the lips interfere with one another. Now to make them kiss. Take it slowly. The shoulder plane cuts finer, but you can see better with the paring chisel. Besides, shearing diagonal grain in hard wood is a rare pleasure. Push the chisel along, bevel up, shearing off the long corner and ends of the square lip until you have only a gentle rise in the middle to flatten with a few strokes of the plane.

Working carefully, you can make the joint close to a perfect miter leaving no trace of the workmanship inside.

And that's the problem.

With the workmanship hidden, the secret miter dovetail joint was the hidden place where piecework cabinetmakers could cut corners—badly. These sweatshops have moved on to other parts of the world, but they are still part of the legacy of the old tools and antique furniture that we so admire. When you see an old piece of furniture disassembled for repair, these hidden joints speak the truth.

Trammels and the Ellipse

When the arena for displaying the cabinetmaker's mastery of wood is an oval, out comes the ellipsograph. There's no doubt about what an ellipsograph does, but how it does it—that takes some thinking. The device is also called a trammel, a bit confusing since trammel points are the spiked heads used to make up a beam compass. These look like pieces from an ellipsograph that lost their X. They just make circles.

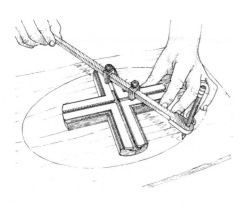

The adjustable trammel or ellipsograph.

You can also generate ellipses with two nails and a string. Draw a long axis line down the center of the length of the tabletop and a short axis line across the width. Measure half of the long axis, then measure the same distance from an outside edge of the short axis to make intersections with the long axis. Tack nails into both of these intersections and into the outside edge of the short axis. Now tie a snug string around these three points and then pull out the outermost nail. Set a pencil within this loop of string and bring it around to trace a perfect ellipse within the rectangle.

The ellipsograph and the nail-and-string method can generate ovals of any size, but the latter is better suited to larger tabletops.

Rule Joint

When the oval table gets large enough to merit drop leaves, we return to circles. Circular rule joints on drop leaves look good even when the leaves are down, and they keep the leaves fully supported when up. Like their namesake joints on folding rules, one element of the rule joint is concave and the other convex—both elements sharing a common center. With a hinge also set at this common center, the drop leaf can fold down with its concave edge constantly cupping close to the convex edge on the main top.

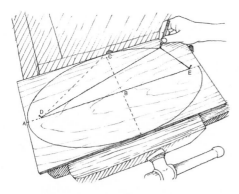

Find distance A, B. From point C, measure this distance to find D and E. A string tied to fit around tacks placed at C, D, and E will describe the oval.

Rule joints use special hinges made with one leaf longer than the other and with the countersunk holes on the face opposite the barrel. Their thickness and individual geometry determine the center of rotation, so have these hinges in hand and see that they are all well matched before cutting any wood. Working from them, and the thickness of the tabletop, you can make scratch stock scrapers that will guarantee a close-fitting joint. I'll describe the steps for drawing the joint first on paper to then use as a guide for filing the steel scrapers. I'll describe this for flush hinges, but you can have them countersunk or, even better, set in sloping recesses.

Draw two parallel lines separated by the thickness of the tabletop, adding one vertical line representing the two butting edges of the top and leaf. Take the hinge and measure the distance from the face of the hinge to the center of its pin. Measure up the same distance from the lower line and draw there a faint parallel line.

Measure down from the top line to the depth of the square-butting fillet at the top of the joint, less than a third of the thickness of the top. Draw another faint parallel line at this level.

The distance between these two faint lines is the radius of your rule joint. If you have a set of hollow and round or table planes, you will want to adjust the depth of the top fillet to make this radius match. Lacking such, the scrapers can do the whole job. Set the point of your pencil compass at the intersection of the vertical line and the lower faint line. Adjust the compass to touch the upper faint line and draw the quarter circle into the leaf half of the joint, then faintly mark the intersection A on the opposite half.

Reset the compass to pivot on point A and draw the arc of the convex element. With a few more straight lines for the fillet and bottom, you have the paper patterns to glue on your scrapers for grinding and filing. The concave element on the leaf needs to be just a shade larger than the convex element, so take this into account. Since the convex scraper shapes the concave element, leave it a pencil line larger than your drawing when you file it to shape.

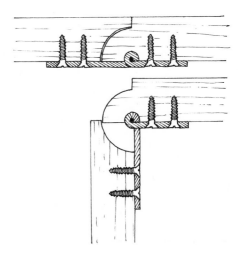

The drop-leaf rule joint never opens up a gap.

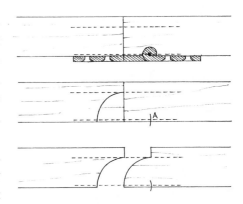

Lay out the arc from the center of the hinge pin.

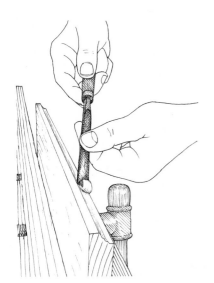

LEFT: *Gauge the guidelines and rough in the hollow with a gouge.*

RIGHT: *Smooth the gouge work with a round plane.*

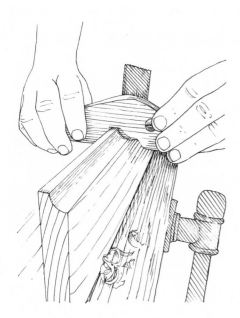

Finish the cove with a scraper mounted in a scratch stock.

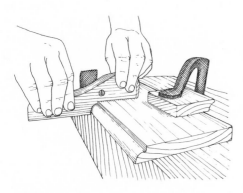

Cut the shoulder with a fillister plane, then chisel and plane the quarter round, finishing with a scraper.

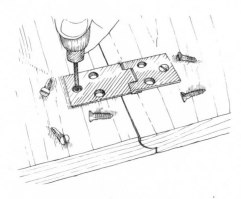

Make holes for the screws with a bradawl after countersinking the barrel of the hinge.

Scrapers can do the final shaping, but faster-cutting tools can remove most of the wood. Lay out the same lines as before on the tabletop and the leaf. When you gauge the shoulder of the fillet on the upper face of the main top, gauge this same line on the underside. You will need this lower line when mounting the hinges.

For the concave element, mark the upper and lower limits of the curve with the marking gauge. You can rough in the hollow with a gouge, or with the fillister or rabbet plane. Follow with a round plane and then the scraper mounted in a scratch stock. The scratch stock can be any handy scrap wood, but it must have shoulders to ride against the guiding surfaces of the tabletop, keeping it precisely in place as you pull it along.

Begin the convex element by sinking the top fillet shoulder with the fillister or rabbet plane. Take off the corner with a plane. Follow with a hollow plane and then use the scraper.

All the care put into shaping the wood won't help if the hinges aren't set right. Assemble the joint with the two elements separated by a piece of paper to assure clearance. Set a hinge across the joint with the long leaf of the hinge on the drop-leaf side. Align the axis of the hinge on the line showing the center of the circle. Poke four little marks at the corners of the barrel of each hinge. Connect these four points and take out the waste wood with a chisel.

If you are countersinking the entire hinge, take care not to cut any deeper than you need, for the error will appear every time you lower the drop leaf. Cut the inset for the hinge at a slope on the drop-leaf side. This leaves the hinge countersunk at the barrel and sloping to the surface at the end—preventing these unsightly gaps.

When the hollow for the barrel is completed, set the hinge in place to mark the screw holes. In any hinging operation, it's wise to use just one screw on each side so you can test the operation of the joint before continuing with the rest of the screws and the other hinges.

Knuckle Joint

What supports a drop leaf when it's up? Sliding supports can pull out, and wings and gate legs can open on metal hinges. But when you want a single rail and leg to swing out, you use the knuckle joint—a wooden hinge with intermeshed wooden barrels turning on a metal pin. This joint shares characteristics of the rule joint, in that each cylindrical knuckle fits in a concave socket on the adjoining piece. Like dovetails, however, knuckle joints always join the ends of boards, not the sides.

You can cut the knuckle joint in the middle of a single board—just remember that you will lose some length to the joint. Both ends must be true and square. Set the gauge to the thickness of the wood and mark all around the ends of both pieces. This defines a long square space, within which we'll shape the cylinder of the knuckles.

If you want a knuckle joint that swings 180 degrees, the cylinder diameter can equal the thickness of the wood. Oddly, a lesser, 90-degree swing requires a smaller cylinder so there will be wood for a stop when the rail is straight. Usually there is some other piece in the table that acts as the stop, so the full cylinder is most common.

The center of this cylinder is the intersection of diagonal lines connecting the corners of the squares on the edges. Confirm this center with your gauge set to half the thickness of the wood, gauging in from three sides to define the same point. Set a small compass at the centers and inscribe the ends of the cylinders on both pieces.

Before we shape the cylinders, we need to saw the line defining the stops. These stops are miters, bevels that meet when the joint reaches a right angle. The line of these mitered faces is the same as the diagonals we drew to find the centers of the cylinders. Take the try square and bring the intersection of the diagonal up and across the faces. These are the first lines to saw, shallow cross-grain cuts reaching to the intersection of the cylinders and the stops.

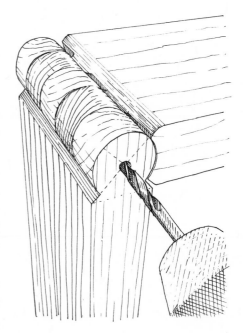

Boring the hole for the pin in a completed knuckle joint.

Now you can shape the cylinders and the stops. As with the secret miter dovetail, rough in the shapes with cross-grain paring and follow with a low-angle shoulder plane. If you want to make a concave sanding block to finish the rounding, take an auger the same size as the wood is thick (or the next larger one) and bore a long hole through a piece of scrap. Saw this in half and you have it.

For the knuckles, divide the width of the rail into five equal parts using a diagonal ruler. Don't bring these four diagonal marks over to the line; just use them to set the marking gauge for strokes delineating the knuckles. Mark both pieces all around both sides before changing the setting, and be sure to run the fence of the gauge along the same edge. Mark Xs on alternating segments to show the waste on both pieces.

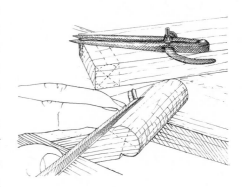

Saw the shoulders of the barrel and pare it round with chisel and plane.

Set one of the ends upright in the vise and confirm the waste pieces. Saw on the waste side of the lines, the edge of the kerf running down the center of the gauged lines. Saw only as deep as the cross-grain cuts defining the shoulders. Remove the waste with chisels as in dovetailing. An in-cannel gouge can round the hollow of the end sockets. On the intermediate sockets, undercut with scooping strokes of the chisel.

When the two fit together and work freely in your hands, clamp them together and drill the pin hole squarely through from both ends, meeting in the middle. Rub some tallow on the surfaces before you drive home the pin.

There is an old belief that whistling while cutting wood for a wagon wheel will cause the finished wheel to squeak. You can extend this whistling prohibition to include the work time on all moving constructs, including rocking chairs and knuckle joints. It's a useful belief, because repeating it is the only polite way yet found to get habitual whistlers in the shop to give it a rest.

Mark each division on both pieces before resetting the gauge. Saw and chisel away the waste.

Toothing Plane and Cabinet Scraper

Speaking of abrasive, I have mentioned sanding only once—not for finishing, but for the final rounding of the knuckle joint. Abrasive paper uses the sharp edges of crushed stone to scrape away tiny bits of wood. A fractured flint or a

A toothing plane in the German pattern.

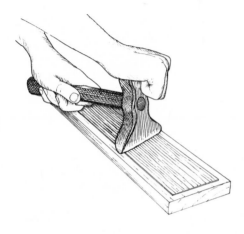

The veneer hammer squeezes the veneer tight to the ground.

Polish the sides of the scraper with the back of a gouge or a steel burnisher.

shard of glass gives a broader sharp edge to scrape with but functions in the same way. I plan to finish off with cabinet scrapers, but there's another tool on the cabinetmaker's bench that also scrapes, the toothing plane.

The grooved iron of the toothing plane sits almost vertically in the body. The grooves on the flat side of the iron become sharp teeth where the bevel side meets them at the edge. The irons come in coarse, medium, and fine, and the planes find work in veneering and in finishing impossible grain.

Even on the mellowest grained veneer, the toothing plane helps the glue hold. The coarse teeth leave tiny channels in both the background and the veneer to ensure that the trapped air can escape and that the glue is not all squeezed out by the veneer hammer—which just looks like one. It's really a wooden squeegee worked along the freshly glued veneer, pushing it flat and squeezing it tight.

And when wood isn't mellow, when it's devil-twisted, passion-flamed stuff that glares back at you, then the toothing plane helps again. In such stuff, solid or veneer, a fine toothing plane may be the only tool that can bring it down without tearing it up. Working the finely set toothing plane in crosshatched patterns softly shreds the wood, leaving it level enough for finishing with the cabinet scraper.

Broken saw blades often live on as scrapers, but a purpose-made cabinet scraper has just the right degree of ductility to let you draw out and turn over a good hooked cutting edge. The words "draw" and "turn" are deceptive, because you create the hook by pushing the steel into shape with a rounded burnisher of even harder steel. This burnisher can be purpose-made or the polished back of a gouge, just as long as it is hard and somewhat cylindrical. Cylindrical is the trick. This shape concentrates all the strength of your arm into one tiny intersection with the scraper—a point of extreme pressure that you draw along the edge to reshape the steel as you need.

Once a scraper starts to scrape instead of shave, it needs sharpening. Sharpening the scraper takes three steps: truing the edge to 90 degrees, drawing out the burr, and turning the burr.

Clamp the scraper in the vise and draw-file the edge with a fine, flat file held squarely and pulled down its length. Follow the file with a whetstone, again keeping everything square. I find it easier to keep the scraper and the whetstone square if I leave the scraper in the vise and pull the whetstone down its length—just as when working with the file.

Now lay the scraper flat on the bench top with the edge even with the edge of the bench. Hold the scraper steady with one hand as you stroke down the very margin of the flat face with the burnisher—polishing and pushing the steel outward into a tiny burr with about eight or ten firm strokes. Give the burnisher a few licks with your tongue as you work.

Slide the scraper over the edge of the bench top so that you can now hold the burnisher vertically and stroke it down the narrow edge of the scraper to turn the burr upward. This takes about five or six strokes, as you gradually increase the angle and reduce the pressure, finishing the hooked edge with the final strokes.

You can't see the cutting edge, but you can feel it with your fingers. More important, you can see it work. Hold the scraper in your fingers, springing it forward with your thumbs. Tilt it forward on the work and push it along. If you have to tilt it uncomfortably far forward before it bites, you've turned the edge a bit too much. The cabinet scraper gets very hot as you work. Walter Rose remembered the workmen's thumbs turning red as raw meat from the heat of their work.

The same principle of the hooked cutting edge can apply to scrapers of any profile, from simple curves to compound moldings. Just remember that tilting a curved scraper forward changes its geometry, flattening its profile as you tilt. A bevel-sharpened scraper (sharpened like a chisel) will work when held at right angles to a surface, but it really is scraping and not cutting a fine shaving like a good hooked edge. The edge of a sharp scraper will give you a fine finish that may be diminished, rather than enhanced, by sanding.

Sandpaper has been around for a century or more, but so have other abrasives such as scouring rush, or Dutch reed. For thousands of years, artisans of all sorts have used sharkskin for fine finishing. Some friends managed to get some skin from dogfish, small sharks, and recommended it to me. They gave me some to try, and I stroked it lightly across my cheek to see just how sharp it was. Before I could try it on the wood, I was stopped by blood dripping on the bench. My entire cheek had been painlessly but thoroughly lacerated by a few million years of evolution.

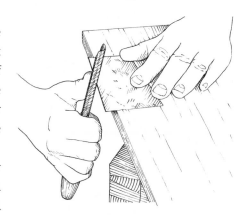

Turn the edge, giving a slightly greater tilt to the burnisher with each stroke.

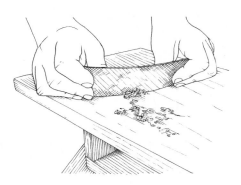

Lean the scraper forward and spring it into an arc as you push it along.

Conclusion A Great Wheel

One day in Williamsburg, a message reached me at the carpenter's yard that the cabinet shop was shorthanded and needed help with something. I figured it was some high-end task like dovetailing a chest or shaving a score of cabriole legs, so I dropped my adze, told the crew where I was going, and headed down to the cabinet shop over the creek.

I eased into the visitor-crowded shop and waited for a couple of Japanese honeymooners to get their photograph. I shivered, chilled as my sweat-soaked colonial outfit cooled in the air-conditioned shop. Along the wall, I spied the work ahead—a great stack of heavy walnut spindle blanks, waiting by the lathe.

Well, I'm a pretty good turner, so I happily worked my way through the visitors toward the lathe and the wall-mounted rack of polished gouges. Then, I noticed that David, one of the journeymen in the shop, was already standing at the lathe and was nodding toward the great wheel that drove it.

Oh, fair enough, I thought. I stepped over the rope, set my hands on the drive handle of the six-foot diameter wheel and began to crank. Slowly, slowly, I worked it up to speed. The first ten minutes of turning were fine. I watched the flapping leather drive belt as it poured from the top of the wheel over my head, ran down along the wall to the wear-polished headstock pulley, and then came rushing back toward my feet. I counted the spindles in the stack. I watched the faint drift of brown shavings feathering to the floor.

I was craning my neck to see if there was any water in a mug sitting on the windowsill when I felt a rap on my shoulder. I turned around to see one of the visitors ducking under the rope barrier. This was easy for him, because he was a little guy, old as my dad, but moving fast. I tried to form the words explaining that the ropes were there for his protection, but I got only as far as taking in air before he pushed me away and grabbed the crank of the great wheel himself.

"Outta the way, kid," he said in some New York accent. "Here's d'way ya do it!"

I reached to direct him back across the rope into visitor world, but the room was transforming. I pulled back. Gone was the clattering roll of the wheel and the lazy cut of the turning. Now the great wheel was flying, the floor hummed along with it at some high-energy harmonic, David struggled to hang on to the gouge as a ribbon of sheared walnut hosed over his shoulder.

"Dat's how ya do it kid; ya gotta put yer ass into it!" he shouted.

I knew this couldn't last, but he kept it up, grinning and shaking his head, never flagging. He was still grinning as I reached out to him again. He grabbed me first and now had me turning the wheel as he ducked back under the rope.

He stood there grinning at me. "All right now, kid. Dat's it! Just put yer ass into it!"

I had no idea what he meant, but at that moment I felt the belt slacken as David parted off the finished walnut spindle. I let the wheel slow.

"Never thought I'd do that again!" he said.

"What . . . ?" was all I got out.

"Ah. I was in the ball turret of a B-17 durin' the war. Two-six-two got us and next thing I know I'm in a parachute. I wake up with some Austrian farmer pokin' me with his pitchfork. Marches me back to his farm and puts me right to work. For six months, 'til the end of the war, I'm turning this wheel for this Austrian farmer." He reached over and rapped my shoulder hard with his knuckle in the way only old guys know. "And the one thing I learned was, ya gotta put yer ass into it."

As he faded back into the shuffling stream, I tried to re-establish boundaries by cranking up my historical interpretation. "The great wheel that you see here is just one of . . ." Across the room I saw his head poke around to listen to me. I stopped short. Here's this guy—one minute he's flying along, next minute a jet plane shoots him out of the sky, next minute he's cranking a great wheel on some Alpine farm in a scene out of the Middle Ages . . .

And I'm trying to teach *him* history.

Appendix

Plan A. Making Wooden Screws

If you look on the back of a nickel, below the dome on Monticello you'll see a pediment with an arched window. This window is in the little attic room where Thomas Jefferson kept his woodworking bench. Here, Jefferson could chisel and plane, pausing now and then to sweep the shavings off of his leather-bound volumes of André Roubo's *L'Art du Menuisier*—The Art of the Joiner.

The dome on Monticello was a Roubo contribution as well—not from his books, but from one of his actual buildings. During a visit to France, Jefferson was inspired to rapturous verse by André Roubo's great wood-and-glass dome spanning the courtyard of the Paris wheat exchange. Under this dome, Jefferson met his lady-love Maria Cosway, and on his return to Virginia, he modeled the dome at Monticello after Roubo's creation. So whatever else we owe him, without André-Jacob Roubo, the back of our American nickel might be blank.

Roubo was the real thing. The son of an impoverished cabinetmaker, he was indentured at age twelve, in 1751, as a heavy construction laborer. In spite of this hard start, he somehow managed to impress Jacques-François Blondel, writer, teacher, and architect for Louis XV. Studies with Blondel gave voice to Roubo's mechanical genius. In 1769 the Académie des Sciences published Roubo's great multivolume *L'Art du Menuisier*, from which I translate his plan for a tap and screw box—*tareau et filière en bois* (part 3, section 3, chapter 13, plate 311).

I have taken considerable liberties with the translation, moving paragraphs and sentences about, as well as using equivalent rather than literal meanings. It's a technical treatise, but there are still echoes of Roubo's childhood poverty in the opening lines. And only a man who served his time as a ragged laborer would feel the need to remind us that "neither the outer form nor the ornamentation of the screw box adds anything to the excellence of these tools."

If joiners had mandrel lathes (or rather, if they could afford them), they wouldn't need screw boxes and taps. But, when they need to make items that can be disassembled (and they must avoid expensive ironwork), then they must have some way to make wooden screws and nuts. There are taps and, therefore, screwboxes of all sizes, from 1/4 inch in diameter, right up to an inch, and even above. You may buy them ready-made at the Marchands Clincaillers; however, it is quite easy for joiners to make them.

Make the tap first. The tap shown in figure 4 is a tool of steel with a wooden handle in the same form as an auger. The lower part of the tool is cut in the form of a screw. The threads project from the hub of the body and are interrupted by four file cuts parallel to the axis of the tap. Figure 2 represents the four projecting angles, which are all perpendicular to the

Pl. 311.

TARAU ET FILIERE EN BOIS, PROPRES AUX EBENISTES.

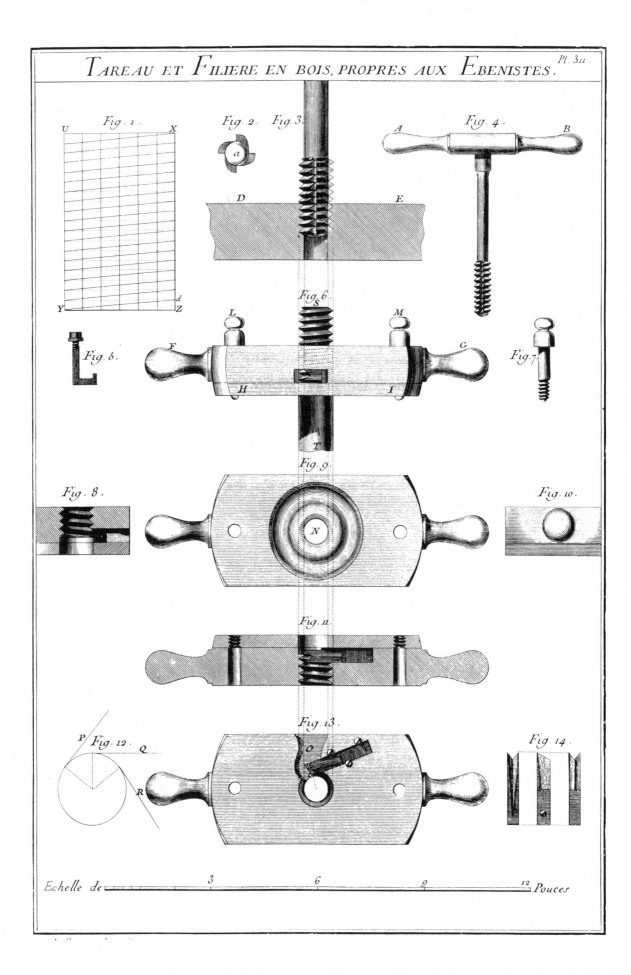

Echelle de ⟶ 3 ⟶ 6 ⟶ 9 ⟶ 12 Pouces

center *a*. When the tap is used, these flats across the screw threads act as cutters to slice and carry away any wood they encounter.

To make the tap, start with a piece of round, forged steel, on the lower part of which you make the screw steps with a triangular file, after drawing them in the following manner.

Draw on a piece of paper a parallelogram, *U, X, Y, Z*, figure 1, of which the width is equal to the circumference of the tap. Divide this width into six equal parts, as the perpendicular lines of this figure indicate; then take half of one of these divisions which you carry from *Z* to *d*, which gives the inclination of the steps of the screw, which, to be good, must be 1/12 of the circumference. As to the pitch (the distance from one screw step to the other), it must not be greater than the distance *d-Z*, that is to say, 1/12 the circumference of the screw, or 1/18 at the least—especially for wooden screws.

When you have drawn the line of one screw step (*Y-d*), add as many parallel oblique lines as you want to make screw-steps on the tap. This being done, glue the paper around the lower part of the tap, observing that the horizontal lines meet perfectly. When the glue is dry, take a triangular file and cut the spaces between each oblique line around the circumference of the tap, right up to the threads indicated by these lines, coming to a sharp edge. If performed right, it is certain that you will succeed to make taps of all sorts of thicknesses as perfect as they can be—without being made on the lathe or with the iron-cutting die.

When you want to use the tap, bore a hole equal in diameter to the tap at the bottom of the screw threads. Put the tap in this hole and turn it from left to right, taking care to withdraw it and lubricate it from time to time. This prevents the wood from jamming and the thread from splitting. The pressure of the tap may even split the piece, which you can avoid by keeping the piece in a vise as you work. To get it started, it helps if the tap is a little smaller at the bottom. Figure 3 shows a cut-away view of a tap that is halfway through a piece of wood.

Having made the tap, it is very easy to make the screw box. This must be made with very dry, hard wood such as boxwood, service tree (*Pyrus sorbus*), etc. The screw box is composed of two main parts; the first of these is the die, which is a piece of wood about an inch thick, three inches wide, and nine or ten inches long, including the two handles *F, G*. On this die is the second part, a piece of wood *H, I*, figure 6, about 3/8 inches thick called the guide plate, which is fastened to the die with two screws *L, M*, which pass through the die and thread into the guide plate as in figures 7 and 11.

Bore a hole *N*, figure 9, in the middle of the die and thread it with the tap as in figure 3. This makes the inside of this hole a nut, with which, along with the V-cutter placed inside the die, one can make wooden screws, as I will explain hereafter.

This V-cutter, shown in figure 14, must be a bit thicker than the height of one of the threads of the screw. The inside of this cutter must be filed out and sharpened keenly at its tip, so that it can cut the wood and not compress it, which will happen, of course, if it jams like the tap. Very often the wood, especially soft wood, compresses instead of being cut, and the threads break.

The edges of the V-cutter must not be made perpendicular to its face, but rather a little inclined ahead of *b*, figure 13, so that it doesn't cut the wood squarely, but a little obliquely, beginning with the outside of the cylinder in which one wants to make a thread. This facilitates the evacuation of shavings, and at the same time, prevents the wood from chipping, seeing that the extremity of the threads is always cut first.

The cutter must be laid out so that its cut, taken on the line of *a*, *b*, figure 13 (which is one of the radii of the circle that forms the threads of the nut), forms at its exterior an equilateral triangle, of which the apex is at point *b*. The V-cutter of the screw box is set in a mortise made in the thickness of the die, with care taken that its point *b*, figure 13, precisely touches the interior circle of the thread and aligns with the internal thread projecting to this same point, as seen in figure 11. This is very easy to do, since it is only a question of making the groove in which the cutter is placed a little more or less deep.

Whether the cutter is inclined as in figure 13 or parallel with the side of the die makes no difference, provided that it is always tangential with the inner circle of the threads, that is to say, that the length of the cutter is perpendicular with the radius of the circle taken at its meeting point, as I have expressed in figure 12, where the lines *P*, *Q*, *R* form right angles with a radius of the circle, to which these lines are tangents.

The cutter must be placed very accurately and strongly supported in its mortise—as much on its width as its depth, and especially at its extremity *c*, figure 13, so that when it is used, it cannot push back, however strong be the pressure of the wood of the screw. The V-cutter can be fastened with two or three screws placed on the two sides, as in figure 13. It is even better to put a hook like that shown in figure 5, which passes through the thickness of the die and is fastened with a nut to clasp the cutter in its width. You might even make a groove on the thickness of the cutter in which the hook can fit, just to fasten it in a firm and solid manner.

Behind the cutter, with respect to its cutting edge, make a groove, *O*, figure 13, called the window, which serves as the exit for the shavings. This window must be of a depth equal to that of the groove where the iron is placed, of which it serves as a continuation.

The guide plate of the die is bored with a hole, the center of which must exactly correspond with that of the die, and the diameter of which must be equal to the bottom of the threads of the latter, as can be seen in figure 11. The first wooden screw threads in the die are removed right up to the point of the cutter so that it can catch the cylinder being fed into it by the guide plate. After the guide plate is centered and fastened by the screws *L*, *M*, finish the screw box on the outside. This finishing will bear no sort of difficulty, seeing that neither the outer form nor the ornamentation of the screw box adds anything to the excellence of these tools.

The guide plate keeps the cylinder in place as it is threaded, which is done in the following manner. Begin at first by turning a cylinder of suitable length and size, which you narrow a bit at its end to give it entry. This being done, put the cylinder in a vise and fit the end of the cylinder in the guide plate of the screw box. Hold the screw box in two hands by the handles F, G, and turn from left to right, pressing lightly down until there are several threads made, which are engaged by those of the die, dispensing with the need for any more pressing. On longer screws, be sure to take off the die from time to time and rub the inside with a piece of soap—more to prevent it from getting too warm from the rubbing of the screw than to facilitate its passage. Figure 6 shows a die all put together and seen from the side of the window, with a cylinder S-T, of which the top part is threaded as I have just described.

When it happens that the piece that you are threading comes up to a shoulder, work as close to it as you can, then remove the guide plate from the screw box. Then thread again as far as possible, so that there remains no more than one turn to reach the shoulder. Finish this thread with a chisel, so that the screw is perfect in its length.

The screw box and tap described by Roubo require some metal working, including annealing, filing, hardening, and tempering. The tap has many teeth and will hold up even if made from unhardened mild steel. The V-cutter of the screw box, however, must be hardened tool steel to hold a keen edge. Old files are time-honored sources for tool steel, and a new file works just as well to make the V-cutters. Just shop around for cheap, imported square ones, likely made of old-fashioned carbon steel.

Files are too hard to cut and shape until you soften them. Anneal the file by heating it cherry red and letting it cool very, very slowly. You can now cut off a length with a hacksaw and shape the V with a triangular file. Re-harden the shaped steel by again heating it to cherry red and plunging it into water. This makes the steel so brittle it can snap, but tempering will bring back some of the toughness. Polish the face of the V-cutter on a whetstone so you can see the oxidation colors as you reheat the steel. Slowly heat the end of the cutter away from edge until the polished surface turns straw yellow and then quickly plunge it into water again. Hone the tempered tool to a fine edge with triangular slip stones.

Plan B. The Carpenter's Tap

While Roubo offers advice on the tools for making smaller screws, the following translated excerpt from Bergeron's 1792 *Manuel du Tourneur* (chapter 1, section 3, plate 7) addresses how to make big screws for bench vises and presses using the carpenter's tap—*taraud de charpentier*. Bergeron was the pen name of Louis-Georges-Isaac Salivet, a prominent lawyer and official in the French Ministry of Justice. When he wasn't considering the fate of the miscreants of Paris, he was treadling away at his lathe.

The tap used to make large nuts is called the "Carpenter's Tap," figures 3 and 11, plate 7. It is composed of two principal pieces, the cylinder, figure 4, cut with a spiral, and the nut, figures 1 and 2. In figure 11 the nut is replaced with the press *A*, *B*. We will begin by describing the manner of making these different parts.

Turn a cylinder of sufficient length and width perfectly equal from one end to the other. While the cylinder is still on the lathe, draw a series of pencil lines around the circumference, spaced at the pitch spacing of the threads. Draw lines down the cylinder along its length dividing it into six or eight equal parts.

Points placed on these lines will delineate the path of the thread. Divide each of these spaces along its lengthwise lines in as many parts as you have made divisions down the length of the cylinder. To avoid confusion, mark the points that indicate a full turn of each thread.

Starting at the bottom right end of the cylinder, at the end of one of the lines drawn along its length, use a pencil and a flexible ruler to draw a line up to the angle at the left that forms the first lengthwise subdivision, from there to the second, and so on. When you have made a complete circumference, you will have arrived at the second main division—one complete turn of the screw. Continue right to the end. This will give a very exact screw, which ends about seven inches from the end of the cylinder. On the last three inches on this smooth part, make a square head so you can fit the tap with a lever.

Make a saw from a steel blade, from 1/2 to 3/4 inch wide and five to six inches long. Tooth it very fine and mount it in a wooden back, figure 5, so that no more than a quarter inch sticks out. Fix it there with three or four pins such as those you see in *a*, *b*, *c*, *d*. This handle is just a bar of wood, nine to ten inches long, sawn along its thickness with a rounded handle. With this saw, follow the line marked on the cylinder. The back will serve as a stop to keep the blade from penetrating more than necessary. You can see in figure 4 the course of this screw on the cylinder.

Pierce at *A*, following the diameter of the cylinder, a flat hole capable of containing a cutter shaped like a barley-grain, figure 6, that fits it exactly and is held by a wooden wedge, figure 7. This cutter is ground to an angle of 60 degrees.

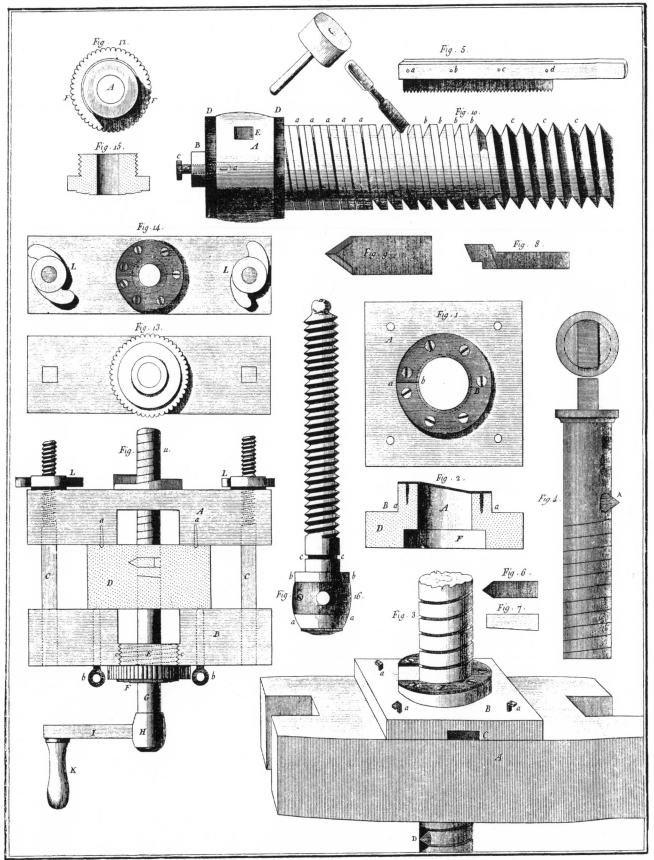

Fig. 12.

Fig. 15.

Fig. 5.

Fig. 10.

Fig. 14.

Fig. 9.

Fig. 8.

Fig. 13.

Fig. 1.

Fig. 4.

Fig. 11.

Fig. 2.

Fig. 6.

Fig. 16.

Fig. 7.

Fig. 3.

For extra strength, you can give the cutter the form shown in figure 8, seen in profile, where you have a reduction in the shank in order not to weaken the cylinder with a large hole. The shoulder you see below the bevel gives you a seat to tap the cutter back in. Instead of leaving the top surface smooth, scrape a hollow there, always protecting the two bevels that form the angle of the cutter. This ensures that the wood is cut instead of being scraped out and crushed as it would be with a plain scraper. By exposing very little iron at a time, with several repetitions, you will make smooth internal threads.

The false nut is made in several ways. We will detail the most common. Take a strong piece of wood, such as cormier (*Pyrus sorbus*, service-tree) or walnut. Make it four inches square for a tap of two inches, five for three, and so on, following the size of the tap. Make the thickness four times the height of the steps of the screw. After you have dressed it with the jointing plane on each face, mount it in the lathe with the universal chuck. Draw two concentric circles in the middle, figure 1. The interior circle's diameter is equal to that of the cylinder that carries the cutter and the exterior about an inch away from that. Cut a hole the exact diameter of the inner circle. Making this hole exactly perpendicular is of the greatest importance. Next, cut away all the wood, A, that is outside the outer circle. Continue thus right up until the projecting part B has at least the height of one step of the screw drawn on the cylinder. Figure 1 shows an end view. Figure 2 shows the cross section, where A is the cylindrical hole, B is the projection, and D is the thickness that you leave raised.

Draw on the exterior of the cylindrical part—of which a, a is the diameter—a thread equally spaced with those on the tap, using the same process you used to draw the screw on the tap. Now, cut the cylindrical part B, exactly following this line in a circular inclined plane, called a "snail," which is made clear in figure 1 and again in figures 2 and 3. Take a piece of sheet iron equal in thickness to the kerf of the saw used to make the screw, and give it the form seen in figure 1. The interior circle in the iron must be 1/4 inch less in diameter than the opening of the hole. The exterior circle of the iron must exactly equal the diameter of the cylindrical part. After having cut out this ring, whether on the lathe or with the file, cut it along its diameter from a to b. Pierce seven holes on this circle for screws from 3/4 to one inch long. Countersink these holes and fix the circle on the inclined plane, A, B. You see by the dotted line, figure 1, that the iron circle exceeds the inner circumference of the hole. It is this excess part that catches the screw traced on the tap.

We now turn to the manner of making a nut with this tap.

Begin by making a hole the size of the cylinder in the piece that you want to thread. Push the cylinder through it with the cutter sunk in its mortise—as you see in B, figure 3, where A is the crossbeam of a press that you want to thread. Fit the false nut B into the thread of the screw and push it flush on the crossbeam. Place it squarely and fix it there with four nails *a*, *a*, *a*, the fourth not being in view here. To make them easier to withdraw, grease them with a little tallow.

Advance the cylinder by turning it to the right, and when the cutter has nearly entered the nut, knock the end opposite the point with the face of a hammer and a piece of flat iron until the point sticks out ever so little from the surface of the cylinder. Then, putting a lever on the head of the cylinder, turn it from left to right, feeling the cutter passing through the wood. When you reach the other side of the piece of wood (which you will easily notice by the end of resistance) take care to empty the shavings from the groove *C* that connects with the hollow *F* that you made beneath the plate *B*. This hollow has a somewhat greater diameter than that of the screw. Turn the cylinder backward until the cutter exits from part *A*. Drive the cutter a little more, and turn it again through the nut, repeating until it is the proper size for the screw.

After having repeatedly run it through, set a bit deeper each time, judge by the length of the protruding cutter if the thread of the nut is deep enough. Withdraw the cylinder entirely and see if the screw, which must have been made first, enters easily enough. If it is wanting a little, put the cylinder back in the false nut and drive the cutter deeper, taking only a very little wood at a time so that the nut will be clean and smooth. It is good that the screw be more tight than slack in its nut, seeing that the wood, no matter how dry it is, always contracts a bit. Rub the screw with a bit of soap and not grease.

Now to make the screw. If the screw does not exceed three inches in diameter, make it with the ordinary screw box. If larger, use the following methods. Turn a big cylinder, figure 10, with a stout head *A* and a spindle *B*. If the screw must meet with great force, as that of a heavy press, furnish the head with two iron bands *D*, *D*, so that the lever passed through the mortises *E*, *E* will not make it split. Pierce a hole in the center of the head suitable to receive an iron bolt *C* on which the head and the collar are turned. The body should be square with a square hole to receive a key, *d*, which keeps it tight in place.

Divide the cylinder in eight, or better yet, in twelve parts, with lengthwise lines. The more the division is multiplied, the more accurately you can lay out the thread. Determine how many turns the screw must make over a certain distance, carefully matching the nut, and mark these divisions down the cylinder. Subdivide each of these divisions into as many parts as you have put lengthwise lines on the circumference, and with a chalk or graphite, trace the thread in the manner that we have detailed above.

Now, trace a second helix in the middle of each turn of the first helix, which will seem to double the screw. If you fear making a mistake, make one line in black and one line in red. Now use a saw in a handle like that in figure 5, or indeed an ordinary saw, but the first is more sure. Set the blade to project all the depth that the thread of the screw must have.

Saw a line to its full depth exactly following the black or red line, as you have determined.

When this saw cut is made along the length of the screw as you see in *a, a, a, a, a*, take a mallet with a well sharpened firmer chisel, and cut down, little by little, all the wood, beginning about 1/8 inch from the line of the saw cut, following a slope, seen in *b, b, b, b*. When you have thus rough-hewn the thread on one side, do as much on the other; next, with a chisel that cuts very keenly, finish giving the thread the form it should have, cutting from the line to the bottom of the saw cut, taking care that you don't leave chisel marks. This will give the screw the form you see in *c, c, c*. To finish it off, pass a half-round rasp over all its surfaces to remove flats of the chisel that always remain. This operation is done with the cylinder lying on a bench, where it is held by its weight if it is heavy, and by a holdfast if not. It is not necessary to follow a thread to the end, endlessly turning the cylinder. You can give a cut of the firmer chisel to all the threads, on the same point of the circumference, as, for example, here at the summits *a, a, a, a, a, a, a*, etc. Then, turning the cylinder a little, you can hew the next summits that present themselves.

Ordinarily, for the screws of screw jacks, wine presses or large machinery, you take some very sound wild stock, some service tree or unhewn walnut, that is to say, the round log — never the quarter — or, finally, some good elm. Carpenters make all their cylinders with the besaiguë with much art. But they never attain the perfection of those made on the lathe.

We will not give here the description of a wine press. These works of carpentry are not our subject of discussion. We will content ourselves to say that the part in which the screw passes in a strong press is called the sommier and is made as seen in *A*, figure 3. It is assembled in the two uprights by means of the fork at either end.

Tapping by the preceding method, one is obliged to fix the false nut on the part with four nails or screws, leaving unsightly holes. It is also inconvenient to back out the cylinder from its guide. An extremely clever means to overcome both disadvantages was communicated to us by the distinguished artist, Hulot. Two strong crosspieces of wood *A, B*, figure 11, notched in the middle, are assembled by means of two wooden screws *C, C*, which enter square to part *B*, and whose other end, partly tapped, enters holes bored in part *A*. These are retained by means of the two nuts *L, L*, which one sees better in figure 14, which is part *A*, figure 11, seen endwise.

Toward the edge of the notch are small, pointed eye bolts *B, B*. These points stick into the part *D* that one wants to tap. They are used to prevent the tap from shifting during the operation. If you fear that the marks of the points will spoil the part, you can back them off.

The top of part *A* carries a false nut, like that in figure 1, that sits on or is set into the piece. If the nut is not yet to the final size, replace part *E, F*, figure 11, with the cylinder fitted in the guide, after having advanced the cutter a little bit. The cylinder is similar to the previous one, except much longer. Part *B* is pierced with a tapped hole much larger than the cylinder, as one sees at *c, c*. Turn a piece of wood, figure 12, which is also tapped, to enter the threads *c, c*, with a rim *F*, of a larger diameter, which shoulders against part *B*. In the center of this disc is a hole *A*, figure 12, just fitting the cylinder *G*, figure 11. The edges *F, F* are grooved, so that one can more easily screw and unscrew this part according to need. It is represented in cross section in figure 15.

When one wants to test if the screw goes well in its nut, part *F* is unscrewed. The hole it exposes is much larger than the cylinder, and when the cutter is withdrawn, there is room to enter the screw, figure 16, to test it.

Figure 14 is, as we said, part *A*, figure 11, seen from the side. One sees the two nuts *L, L*, and the false nut *M*.

Instead of a lever, one can use a wooden crank *I, K*, figure 11, to get a continuous and more uniform motion. This works only for smaller nuts. It is always necessary to use a lever for heavier parts because of the turning resistance.

Figure 16 is a screw from a carpenter's bench vise or a sawing press. This is the size of the screw for the nut that we are making—part *D* is the leg of a carpenter bench. The head of the screw is fitted with a broad iron hoop, inset well into the head and resting against a shoulder at *a, a*. Turn a good flat shoulder at *b, b*. Fix the hoop on the head screw with four long wood screws, as one sees at *d*.

Bore the head through its diameter to receive the lever, which tightens the jaw of the vise against the bench. To make the front jaw return when one screws backward, make a small groove *c, c* on the screw neck in which one or two keys ride, which pass through the thickness of the front vise jaw. By this means, the screw can turn easily, but when one turns backward, it takes the vise jaw back with it.

Plan C. A Roubo Bench

The common workbench of André Roubo's time was a single massive plank sprouting four legs. I say "sprouting" rather than "supported by" because this form of bench uses "stool" rather than frame construction—more like a Windsor chair than a table. The version I describe here, fitted with a toolbox with a lock-ing lid set between the stretchers, is based on one I saw in the town of Isle sur la Sorgue in southern France. Simple as it is, this is a challenging bench to make. The timbers are big—hard to find and hard to work. The joints are big too, yet require precise fitting. Done right, though, it's a workbench beyond compare.

When trees were big, so were benches. Old benches, with tops six inches thick and two feet wide, could stand stable on perpendicular legs. As avail-able timbers for the top narrowed, however, perpendicular legs would make the bench too tipsy. But, just like a Windsor chair, narrower-topped benches can regain their stability by splaying out their back legs. These splayed back legs solve one problem but present another. The back legs intersect the top at an angle, with their tenons fitting into mortises angled through the top. The front legs, however, are perpendicular. Having the front and back legs come in at different angles poses no assembly problem when they are independent pieces, but here they are joined by the broad stretchers. In a Windsor chair you can assemble the converging pieces—legs, top, and stretchers like a car wreck played backward, driving each joint tighter as you go. You can build this bench that way, but I propose you try a more amusing solution, letting the court jester of joints rescue the old king of benches.

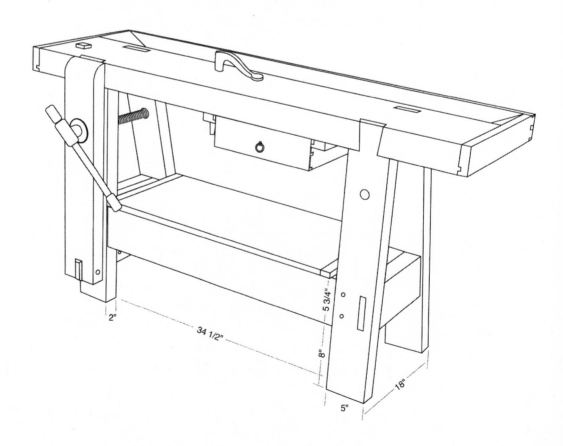

If you look at the dovetails connecting the front legs of the bench to the top, you'll see that they slope in two directions, both up and down and in and out. This means that the legs can never pull down; nor can they pull out to the front. The advantage of this style of dovetail is that it can never come apart. The disadvantage is also apparent — it can never go together.

Or so it seems. This is the rising dovetail joint. It's usually found only in puzzles, fooling us by hiding its slope while letting us continue to think at right angles. The rising dovetail is just a normal dovetail tenon (or pin) tilting toward us. The tenon is at an angle to the length of the leg and to the edge of the bench top. This exposes an oblique slice that appears wider at the end and makes the rising dovetail look impossible to assemble. In our case, however, the rising dovetail makes the bench easy to assemble, because it goes together at the same angle as the sloping back legs. The rising dovetails make it possible to join the bottom frame entirely and then drop the top on with a backward sliding motion.

Start with the top of the top. You want the hardest, thickest, widest piece of wood you can get, but you use what you have. I usually work with oak, but for this bench I lucked onto a plank of hard maple, three inches thick, ten inches wide, and ten feet long. I decided that two, five-foot-long benches would be more useful to me than one longer one. I'm glad I chose as I did, because the five-foot version is almost too much for me to move. The two front legs and front stretcher are also hard maple. The back legs and stretchers are soft maple, which saved money and made the work a little easier.

A flat workbench top is the foundation of everything that you make upon it. In this case, the flat bench top is also the starting point for everything made below it. You level the top, true the front edge, and build on from there. Leveling a great, wide piece of rough-cut hard maple is no joke. You may need a good bit of adze work before using the planes as I describe in Chapter 7. Roubo recommends that you orient the grain of the bench with the heart side up. This ensures that further seasoning will cause the top to crown rather than hollow. Once you establish the flat plane of the top and the perpendicular front edge, you build downward from there to the floor.

The rising dovetail is the second challenge. Our brains struggle to describe three-dimensional work arranged at right angles, and fail utterly when it comes to describing these angled intersections. Even after you've cut one of these joints, you're still not quite sure what you've done. If you have not cut one before, practice on cheaper stock, following the steps illustrated. The dimensions given in the drawing create a slope equal to that of the back leg, making assembly easier.

The back legs have straight tenons with beveled shoulders. They fit into angled mortises, bored and chiseled through the bench top. In this case, the angle of the mortise is easy to find. The top is 10 inches wide, and the desired spread at the feet is 18 inches, so this puts the back legs 8 inches out of plumb. Since the total height from floor to bench top is 32 inches, the splay of the back legs is 8 in 32, or, 1 in 4. Set your bevel to this angle and use it to draw the passage of the mortise on the end grain of the top. Bring these lines across the face and

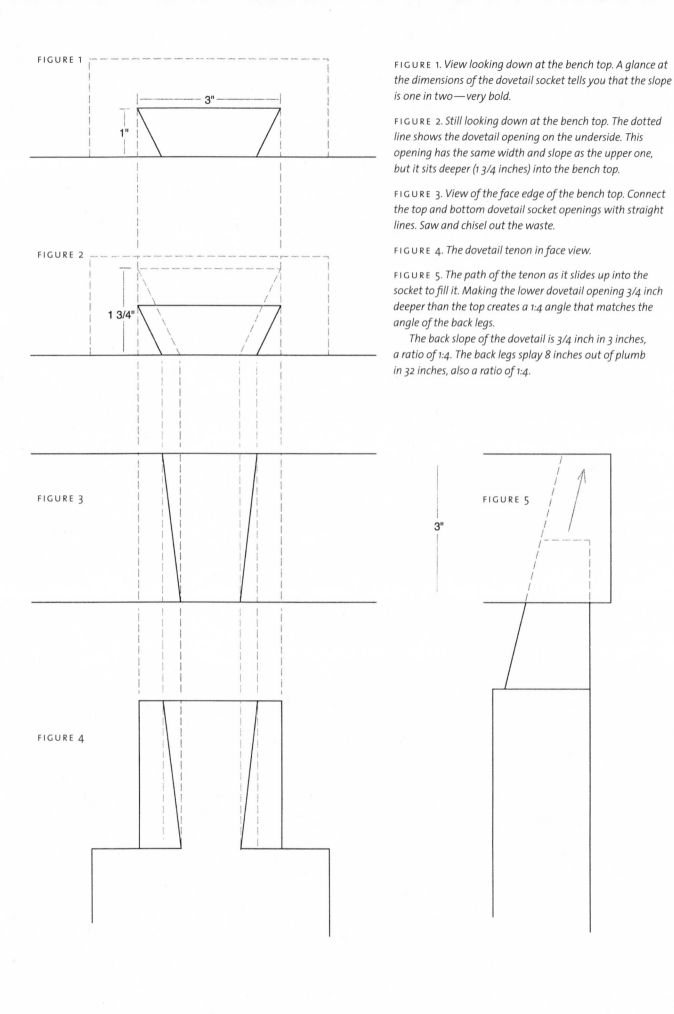

FIGURE 1. *View looking down at the bench top. A glance at the dimensions of the dovetail socket tells you that the slope is one in two—very bold.*

FIGURE 2. *Still looking down at the bench top. The dotted line shows the dovetail opening on the underside. This opening has the same width and slope as the upper one, but it sits deeper (1 3/4 inches) into the bench top.*

FIGURE 3. *View of the face edge of the bench top. Connect the top and bottom dovetail socket openings with straight lines. Saw and chisel out the waste.*

FIGURE 4. *The dovetail tenon in face view.*

FIGURE 5. *The path of the tenon as it slides up into the socket to fill it. Making the lower dovetail opening 3/4 inch deeper than the top creates a 1:4 angle that matches the angle of the back legs.*

The back slope of the dovetail is 3/4 inch in 3 inches, a ratio of 1:4. The back legs splay 8 inches out of plumb in 32 inches, also a ratio of 1:4.

FIGURE 1

3"

1"

FIGURE 2

1 3/4"

FIGURE 3

FIGURE 4

FIGURE 5

3"

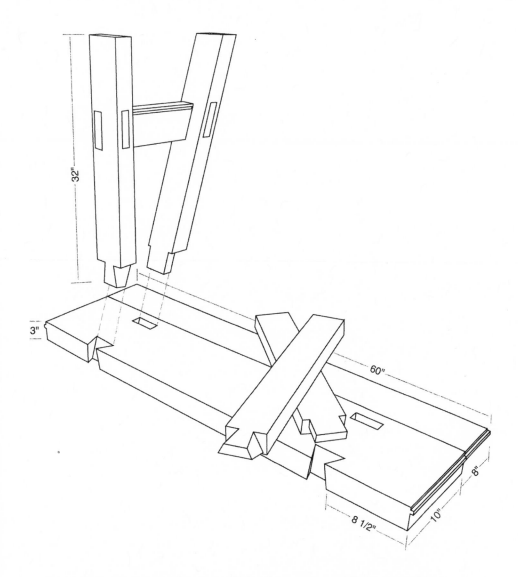

underside to locate the actual mortise. Using the same 1:4 setting on your bevel, use it to guide your auger as you bore down from the top face of the bench.

In spite of all your care in truing the stock and fitting, the front legs may need a slight push or pull to bring them dead perpendicular to the top. The back legs give you something to push or pull against. Fit all the legs as accurately as you can, drive them into place, and bring the front legs square by either pulling together or spreading apart the feet of each pair. With the legs thus wedged or clamped, hold the stock for the stretchers against them and mark them to fit. French workbenches typically have a shelf for tools set in between, or on top of, the stretchers. The stretchers on this bench are wide enough to fit a floor into grooves plowed around their lower edges. Add a lid, and you have a tool chest. A drawer under the top is also handy, and the one in the Roubo illustration is fitted with a lock as well.

Once the legs and stretchers are joined, stand the base upright, set the top on, and start the tenons into the mortises. Rather than pound on the top to drive the joints home, lift and drop alternate ends of the bench on a stout floor, letting the mass of the top drive itself down. The top will move three inches down and 3/4 inch back as the dovetail tenon rises to fill its space.

The solid, level floor required for this mode of assembly also serves when leveling the bench. Shim up the legs to bring the top level, and use a plumb bob and string to check the front legs as well. You can level the top with more planing, but you can't as readily correct the plumb of the front legs. When it looks level, find the leg with the widest gap between it and the floor, set your dividers to this gap, and scribe it around all four legs. Move the bench to a different spot on the floor and repeat the leveling and scribing before you trim the feet to the scribed lines—the floor may not be as level as it looks.

Now that you're up on your legs with that nice workbench top, it seems a shame to lose any part of it. Still, the bench top has to sacrifice a bit of length to create the tongues to support the frame of the tool well. Use a fillister plane followed by a rabbet plane to take down the ends of the top, leaving a 3/8-inch-square tongue to fit into the matching groove plowed into the skirt boards. This groove in the skirt boards continues all the way around the inner face of all three pieces to support the bottom of the tool well. The back edge of the bench top also gets a groove at the same level to support its side of the tool well bottom.

The flat, heavy top is the foundation, but you still need something to hold the wood against it. You can always sit on the wood for mortising, but for planing and sawing, you need a bench stop, a vise, and some holes for the holdfast. Bore and chisel the mortise for the bench stop—the square sectioned block that you tap up and down to catch the ends of boards as you plane them. Get the holdfast in your hands and try its fit in a test hole before you bore any holes in the bench. Augers are graduated in 1/16-inch increments, and making the hole that much too big means forging another, larger holdfast. This form of bench is best suited to take a front leg vise. A two-by-five-inch leg is fine for an iron screw, but not quite stout enough to take a large wooden screw, so *bœuf* it up as you see fit.

Plan D. Hasluck's Bench

Paul Nooncree Hasluck seems to have been the most prolific how-to writer of the last century. You'll find under his name books on making clocks, violins, picture frames, saddles, and motorcycles, to name a few; volumes on working with metal, glass, cordage, bamboo, and wood; and works devoted to photography and rustic carpentry. Born in 1854, by the beginning of the twentieth century Hasluck was editing several how-to magazines and compiling the reader contributions into the many books that bear his name.

I very much like this knock-down workbench from Hasluck's *Handyman's Book*. It's sturdy but easily knocks apart for moving. The vises are simple but can clamp wood both in their jaws and on the bench top. Unlike the Roubo-style bench, this one is constructed like a table. The top is separate and drops down onto the base. The top may be only two inches thick, but it derives some stiffening from the skirt dovetailed around it.

The best wood will make the best bench, but lacking beech, birch, or maple, don't hesitate to make this design in hard pine or oak. I use one with a white oak top, front, and vises. The rest of the wood is red oak, except for the tulip poplar legs. It is rougher than I might like, but it's still a workhorse.

Level the top with the heart of the tree facing upward. True the face side, edges, and ends, but leave the under face rough—it only needs to sit level where it rests on the base. Removing any more wood than that from the underside only makes the top thinner and weaker.

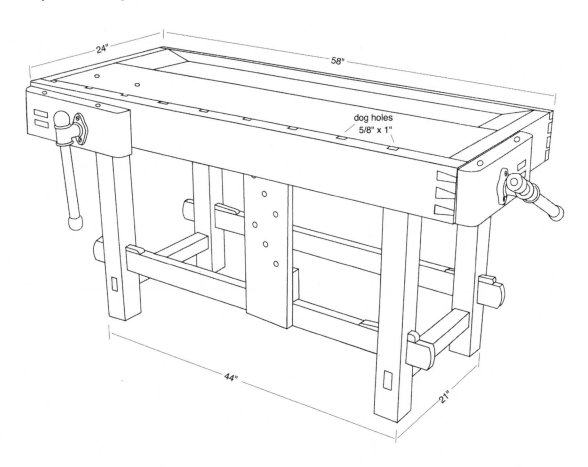

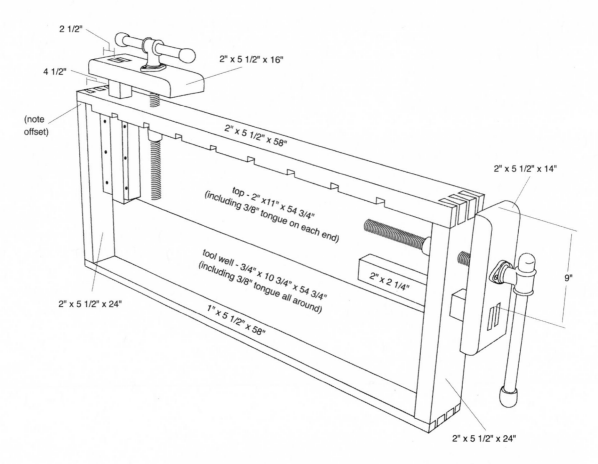

2 1/2"

2" x 5 1/2" x 16"

4 1/2"

(note offset)

2" x 5 1/2" x 58"

2" x 5 1/2" x 14"

top - 2" x11" x 54 3/4"
(including 3/8" tongue on each end)

tool well - 3/4" x 10 3/4" x 54 3/4"
(including 3/8" tongue all around)

2" x 2 1/4"

9"

2" x 5 1/2" x 24"

1" x 5 1/2" x 58"

2" x 5 1/2" x 24"

Just as in the Roubo bench, the skirt framing the tool well hangs on tongues left on the end grain of the thick top. The bottom of the tool well sits in grooves plowed around the inside of the skirt and in the back of the heavy top. In the Roubo-style bench, the bottom of the tool well just sticks out in space. Here, the tool well bottom sits on the top rail of the base, adding stability.

Make the skirt about three times as wide as the top plank is thick. This will give room for the vise screws and the vise arms to fit flush underneath the top. The dovetails joining the corners of the skirt are the same as any other—just larger. The negative spaces are big enough that an auger can help you remove much of the waste. Saw the sides of the sockets down the grain and then bore through the "root" with a brace and bit, leaving just a little wood to take off with the chisel.

This bench can hold a board laid flat on the top, pinched between a peg set in the top of the end vise and a dog set into one of the holes in the front skirt board. These holes are just gently angled dados sawn and chiseled across the back side of the front skirt. The angle of the holes causes the working dog to dig deeper under pressure. You probably don't want to sacrifice any width of either the top or the front by planing a tongue onto one of them to fit into a groove in the other. Pegs, splines, or screws can join the two without losing any width.

Hasluck writes that the double mortise and tenons on the vise jaws and the arms are "troublesome to make" but worth it for the added strength. I agree on both counts. Only the vise on the front needs tracks for the arm, screwed on

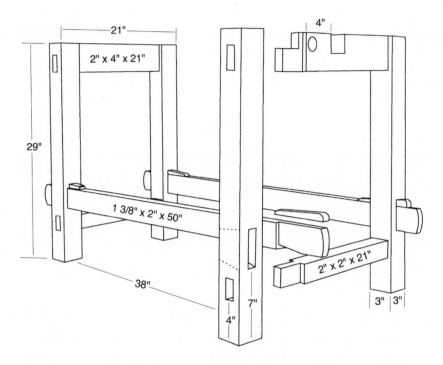

the underside of the top. The arm of the end vise passes through a notch cut into the top rail of the base, giving it plenty of stability. The vises use 1 1/4-inch tail vise screws and are the same ones used when making a leg vise. When the bench top drops down on the base, the top rail of the left-hand frame fits snugly between the vise track and the nut for the screw.

The base is two simple frames joined by long stretchers, all mortise-and-tenoned together. I had intended to use poplar legs only until I could cut some proper oak ones, but the poplar does just fine. You can certainly join the top to the base with screws set through the front and back skirts, but this makes moving the bench more difficult. One person can carry the top by itself (with the vises removed)—but if you connect the top to the legs, you'll need help. The fully assembled base is plenty light enough to carry, but can quickly break down further for packing if you unwedge the long rails. These wedged dovetail tenons are remarkable. Slip the half-dovetailed tenon into the mortise and drive home the tapered wedge to make a complete dovetail and an absolutely stiff connection. It's worth making this bench just to have this joint to admire every time you tap it up.

Plan E. A Spring-Pole Lathe

Perhaps because of its simplicity, the spring-pole lathe is easily dismissed as a make-do device. This lathe, adapted from an old German technical encyclopedia, is no make-do. It's precise, portable, powerful, adjustable, adaptable, and self-contained. The two dead centers permit no play in the workpiece, and the direct drive of the cord wrapped around the wood loses no power in transmission, friction, or vibration. I want you to make this lathe.

I based the entire design on readily available 1 1/2-inch softwood plank. You'll usually find that the best quality wood at the lumberyard is saved for the widest construction timbers, from 2-by-10s on up. Find a tight-grained timber and rip it down to the widths required. The movable right hand puppet is glued up from two pieces of the 1 1/2-inch stock to give it greater body. This sandwiched construction makes it convenient to cut the through mortise for the wedge that holds it to the bed.

Ash and hickory do well for the spring poles, but whatever stiff wood you can find will do—use mop handles if you have to. Fast-grown hickory and oak is stiffer than slow-grown wood with lots of close growth rings, but remember that the opposite is true for pines. The connecting rod between the springs and the rocker arm can be wood or wire, but not something that will stretch and muddy the action. The ring around the two spring rods slides left and right and varies the spring strength from weak to wow! Saw a scrap length of 1/2-inch copper pipe in half down its length and flatten one half into a stout strip. Bend it to fit loosely around the two rods, drill through the overlap, and rivet it shut.

While you are at the hardware store looking for copper rivets, pick up two 1/2-inch hex bolts for the centers. You may also want a 1/2-inch bronze sleeve bearing for the rocker arm pivot. For the axle, you can usually find rods of 1/2-inch mild steel stock, or you can hacksaw a bolt to length. The tool rest has a few screws in it, and needs a carriage bolt, a nut, and some washers to make the adjustable connection to the lathe bed.

You can make centers with threaded shafts to adjust the pinch of the turning wood, but I think you're just as well off with centers fixed into the head and tail stocks. To vary the pinch on the spinning wood, you just tap the movable tail stock with a mallet or with the handle of your turning tool.

For this kind of dead, dead center, you'll first fit the hex bolt into the wood, then remove it, hacksaw off the head, and grind it to a point before screwing it back into place. Because you'll only be able to grip the headless bolt with pliers, the hole needs to be accommodating. Bore the pilot hole in the left-hand upright and thread the hex bolt into it. Back the screw out and file across a few threads to make it cut like a tap. Turn the bolt back in and out a few times—enough to be sure that it will fit in easily but snugly.

Saw the head off and file or grind it to as conical a point as you can by eye. The centers now have to be turned to perfect, polished cones, or they will quickly enlarge their seat in the end grain of the turning piece. I usually set the rough-filed piece in a post drill, a large hand-cranked drill press, and hold files and stones against it as it turns. Alternatively, you might make a V-trough to hold the bolt at an angle to a grindstone. Keep rolling the bolt in the V and you should be able to grind a pretty good point. Screw the completed center into the lathe head, then push the other lathe head against it to mark the place for the second center.

The drive cord always passes down the front of the workpiece, after wrapping twice around it, so that the wood turns toward you from the top as you push down on the treadle. A natural fiber cord frays very quickly from rubbing against itself. Synthetic cords wear better but look out of place. The best cord is round leather belting, such as that used for treadle sewing machines. Spring-pole lathes commonly have proper treadles too, usually a triangular set of sticks with one side hinged to the floor. I have become used to working with just a loose slat for a foot treadle. Whatever you use, make the treadle light enough that you're not fighting its inertia but stiff enough to keep the turning action lively and decisive.

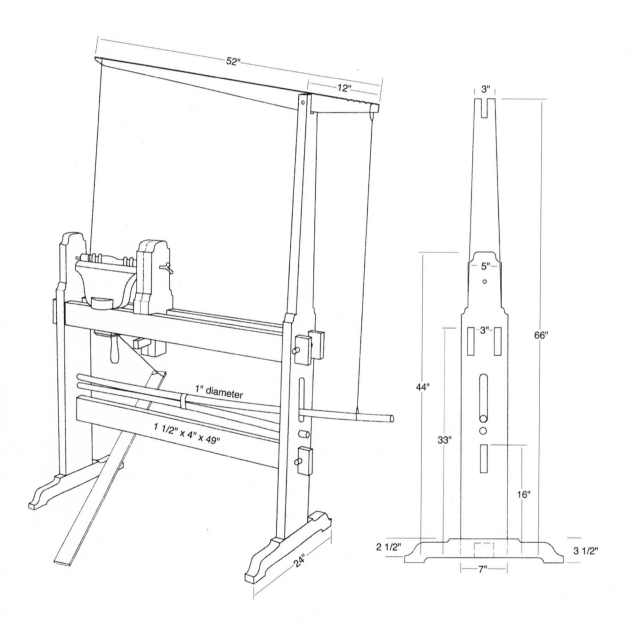

Plan F. A Treadle Lathe

This is a good working lathe that makes up in hardware availability what it might lack in historical accuracy. You can build this lathe with common lumber and hardware store materials, but it does require some heating and hammering to bend the crankshaft and forge the drive center. If you are not set up to do this yourself, it's a fine opportunity to support your local blacksmith.

The lathe frame is just three right triangles connected by four horizontal 2-by-4 rails. Two rails form the bottom, and two serve as the bed of the lathe. The tool rest and the tail stock are the same as on the spring-pole lathe, so I'll concentrate on the moving parts, starting at the foot and working upward.

The treadle is all 1-by-3 stock, cut away to half that width wherever possible to keep it light. The rear pivots on two lag screws set into the frame at both ends. On the left front is the protruding bolt that links to the connecting rod and transmits your foot power to the crank. The cantilevered cross-bracing in the treadle keeps it stiff, even when you're working at the right end—far from the connecting rod. The drawing shows how the diagonal cross brace that runs to the left front passes over the other one in a half-lap.

The connecting rod transmits the treadle action to the crankshaft, which transforms the reciprocating motion into continuous rotary motion. The crankshaft is a 1/2-by-12-inch bolt, sawn off to the finished length after heating and making the two right-angle bends.

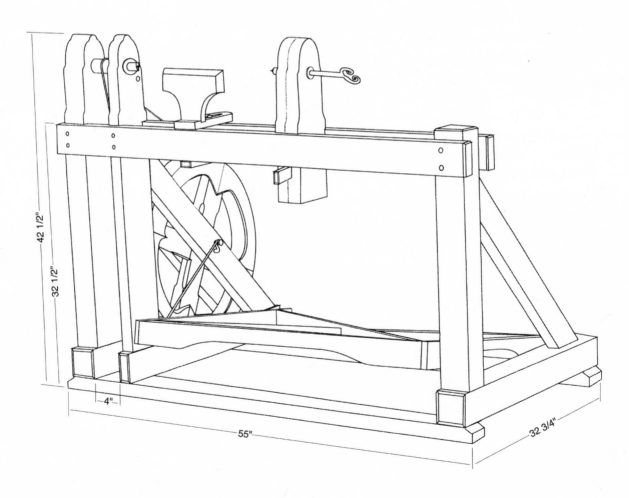

42 1/2"

32 1/2"

4"

55"

32 3/4"

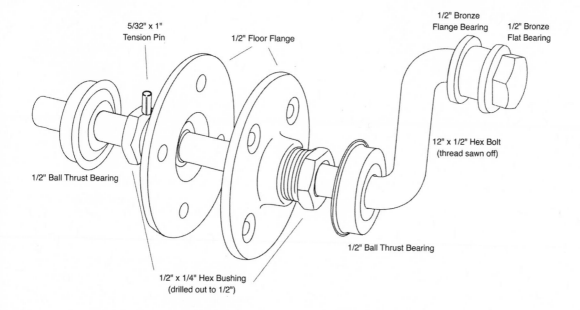

5/32" x 1"
Tension Pin

1/2" Floor Flange

1/2" Bronze
Flange Bearing

1/2" Bronze
Flat Bearing

1/2" Ball Thrust Bearing

12" x 1/2" Hex Bolt
(thread sawn off)

1/2" Ball Thrust Bearing

1/2" x 1/4" Hex Bushing
(drilled out to 1/2")

Before bending the bolt, you may want to slip a bronze flange bearing on the end to ease the friction of the connecting rod. You need to heat the bolt at least to a dull red before bending it, and the bronze sleeve has to go on quickly before it cools. Of course the iron shaft expands as you heat it, so the bearing may be a tight fit. You can heat the area of the bend with the bearing already in place, but remember that bronze bearings are permeated with oil, so don't be surprised when they flame up.

The first bend is just an inch below the bolt head—just enough to give the connecting rod a place to run. The second bend defines the length of the lever—the crank arm. The length of the crank arm helps determine the rise and fall distance of the treadle, but the connecting rod influences this as well. If you want more treadle travel for more torque, link the connecting rod closer to the rear of the treadle. Linking the connecting rod more toward the front of the treadle reduces travel. Try a crank offset of 2 1/2 to 3 inches and adjust the connecting rod link to give you about six to eight inches of treadle action.

The remainder of the 1/2-inch shaft both supports and transmits power to the flywheel. There's a lot of shearing force at play where the shaft connects to the flywheel, but a single steel tension pin passing through the axle and the iron plumbing flange will serve—if the lathe is driven rhythmically and sympathetically. A proper historical treadle lathe uses much more blacksmithing and a lot less time pondering plumbing parts, but this does the job.

The inertial mass of the flywheel smoothes the action and stores energy as it spins, supported by the ball thrust bearings mounted in the frame. The four spokes are just two 1 1/2 inch thick pieces half-lapped at the center. The perimeter of the wheel is a sandwich, glued and nailed together in overlapping quadrants. The filling of the sandwich is made of four pieces of the same thickness as the spokes, butted together at their ends. The bread of the sandwich, the 3/4 inch thick layers on the outside, overlaps the butt joints and holds everything together.

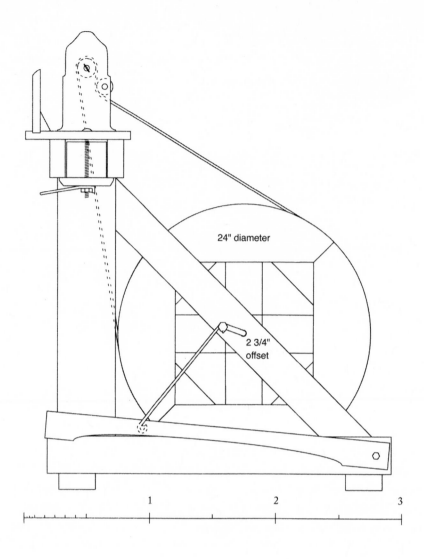

24" diameter

2 3/4"
offset

1   2   3

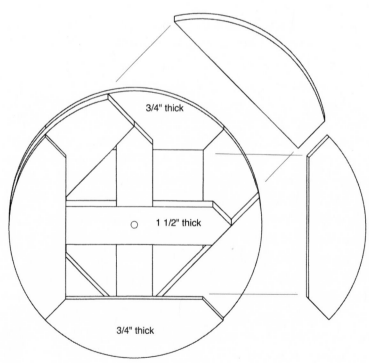

3/4" thick

1 1/2" thick

3/4" thick

A heavy flywheel built in layers.

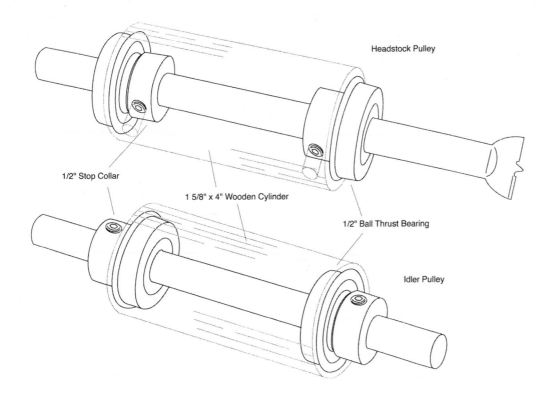

Headstock Pulley

1/2" Stop Collar

1 5/8" x 4" Wooden Cylinder

1/2" Ball Thrust Bearing

Idler Pulley

The flywheel is also the driving pulley for the lathe. The cord transmits its power to the driven pulley in the head stock. For every single turn of the flywheel, the driven pulley has to make about 14 turns, the difference between their diameters accounting for the increase in speed. The difference in diameters also causes a problem—the cord slips on the small pulley. The large driving pulley spreads the cord so much that it contacts only a small area of the driven pulley. There isn't enough grip, and the cord slips.

I originally used flat leather belting on my lathe, so I needed an idler pulley to make the belt run tighter around the driven pulley. This idler is just a wooden shell with bearings stuck in either end. Once, however, when the flat belt came unstitched at a critical moment, I substituted a round cord tied with a square knot—and then I never went back. The cord runs in a groove cut around all the pulleys and is easy to tighten as it stretches—just retie the knot. Using a round cord also offers an alternative solution to the slipping pulley problem—crossing the cord so it runs in an infinity figure. This makes the flywheel turn in the opposite direction from the wood, but that matters not at all. The gripping problem is gone.

In the headstock pulley, power goes from the two-inch diameter wooden shell down to the small shaft. The collars set into the ends of the wooden shell give enough surface contact to prevent slipping. Holes drilled through the wooden shell give access to the set screws in the collars that tighten them to the shaft. The shaft drives the workpiece, pinched between it and the dead center on the tail stock. The pinching grip on the turning workpiece imposes lateral pressure on the shaft, pressure carried by the left-hand collar running against the ball thrust bearing in the head stock.

The headstock shaft poses the final challenge. At one time you could buy lathe drive centers that fit on any 1/2-inch shaft. I have not seen these around for a while, so you will probably need to forge a spade-type drive center. This older style of drive center with two blades was used for centuries and works just as well as the familiar four-bladed version. The hex head of the 1/2-inch bolt used for the headstock shaft will give you enough metal to reforge into a shape resembling a modern spade bit. Leave the central spike oversized so you can true it with a file as it spins in the lathe.

In use, the lathe takes a light touch, with the ankle doing most of the work. Until folks get accustomed to it, most don't lift their foot enough to let the wheel make it all the way around every time. This jerking motion is hard on the crankshaft connections, so let new users take the time to feel the rhythm before they start cutting wood.

Bibliography

Abbott, Mike. *Green Woodwork: Working with Wood the Natural Way.*
    London: Guild of Master Craftsman Production, Ltd., 1989.
Amman, Jost, and Hans Sachs. *The Book of Trades, 1568.* New York: Dover, 1973.
Arnold, James. *The Shell Book of Country Crafts.* London: John Baker
    Publishers, 1968.
Bealer, Alex. *Old Ways of Working Wood.* Barre, Mass.: Barre Publishers, 1972.
Becksvoort, Christian. *In Harmony with Wood.* New York: Van Nostrand
    Reinhold Co., 1983.
Blandford, Percy W. *Country Craft Tools.* New York: Funk & Wagnalls, 1976.
Diderot, Denis, et al. *Encyclopédie.* 17 vols. Paris, 1751–65.
Dunbar, Michael. *Restoring, Tuning, and Using Classic Woodworking Tools.*
    New York: Sterling Publishing Co., 1989.
Eaton, Allen. *Handicrafts of the Southern Highlands.* New York: Russell Sage
    Foundation, 1937.
Edlin, H. L. *Woodland Crafts in Britain.* B. T. Batsford, 1949.
Feller, Paul, and Fernand Tourret. *L'Outil, dialogue de l'homme avec la matière.*
    Brussels: Albert de Visscher, 1969.
Goodman, W. L. *The History of Woodworking Tools.* New York: McKay, 1964.
Goodman, W. L., and Jane and Mark Rees. *British Planemakers from 1700.*
    Mendham, N.J.: The Astragal Press, 1993.
Graham, Frank D. *Carpenter's and Builder's Guide.* 4 vols. New York: Theo.
    Audel & Co., 1923.
Greenhalgh, Richard, ed. *Joinery and Carpentry.* 6 vols. London: The New Era
    Publishing Co., Ltd., n.d.
Griffith, Ira Samuel. *Essentials of Woodworking.* Peoria, Ill.: Manual Arts Press,
    1922.
Hampton, C. W., and E. Clifford. *Planecraft: Hand Planing by Modern Methods.*
    Sheffield, Eng.: C. & J. Hampton, Ltd., 1959.
Hartley, Dorothy. *Lost Country Life.* New York: Pantheon, 1979.
———. *Made in England.* London: Methuen, 1939.
Hasluck, Paul N. *The Handyman's Book.* Berkeley, Calif.: Ten Speed Press, 1987.
Hayward, Charles H. *Cabinet-Making for Beginners.* New York: Sterling
    Publishing Co., 1979.
———. *The Complete Book of Woodwork Joints.* New York: Sterling Publishing
    Co., 1979.
Hazen, Edward. *Popular Technology.* 2 vols. Albany, N.Y.: Early American
    Industries Association, 1981.
Hewett, Cecil A. *English Historic Carpentry.* Fresno, Calif.: Linden Publishing,
    1997.

Hibben, Thomas. *The Carpenter's Tool Chest*. Philadelphia: J. B. Lippincott Company, 1933.

Holtzapffel, Charles. *Turning and Mechanical Manipulation*. 2 vols. London: Holtzapffel, 1875.

Holtzapffel, John Jacob. *Hand or Simple Turning: Principles and Practice*. New York: Dover, 1976.

Hulot, M. *L'art du tourneur mécanicien*. Paris: Roubo, 1775.

Hummel, Charles F. *With Hammer in Hand*. Charlottesville: University Press of Virginia, 1968.

Jones, Bernard E. *The Practical Woodworker*. Berkeley, Calif.: Ten Speed Press, 1983.

———, ed. *The Complete Woodworker*. Berkeley, Calif.: Ten Speed Press, 1980.

Jones, Michael Owen. *Craftsman of the Cumberlands*. Lexington: University Press of Kentucky, 1989.

Jordan, Terry G. *American Log Buildings: An Old World Heritage*. Chapel Hill: University of North Carolina Press, 1985.

Kababian, P. B. *American Woodworking Tools*. Greenwich: New York Graphic Society, 1978.

Klemm, Friedrich. *A History of Western Technology*. New York: Charles Scribner's Sons, 1959.

Knight, Edward H. *Knight's American Mechanical Dictionary*. 3 vols. Boston: Houghton, Osgood, and Co., 1880.

Lamond, Thomas C. *Manufactured and Patented Spokeshaves and Similar Tools*. Lynbrook, N.Y.: published by author, 1997.

Mackie, B. Allan. *Notches of All Kinds*. Pender Island, B.C.: Log House Publishing Company, Ltd., 1990.

Mayes, L. J. *The History of Chairmaking in High Wycombe*. London: Routledge & Kegan Paul, 1960.

Mercer, Henry. *Ancient Carpenter's Tools*. Doylestown, Pa.: Bucks County Historical Society, 1929.

Miller, Warren. *Crosscut Saw Manual*. Missoula, Mont.: U.S. Forest Service, 1978.

Moxon, Joseph. *Mechanick Exercises*. London, 1683.

Murdoch, R. Angus. "Wood Shavings: An Oral History with Robert Simms." 1990 (unpublished).

Nicholson, Peter. *The Mechanic's Companion*. Philadelphia: John Locken, 1849.

Noyes, William. *Handwork in Wood*. Peoria, Ill.: Manual Arts Press, 1912.

Pain, F. *The Practical Wood Turner*. London: Bell and Hyman, 1983.

Panshin, A. J., and Carl de Zeeuw. *Textbook of Wood Technology*. New York: McGraw-Hill, 1964.

Phelps, Hermann. *The Craft of Log Building*. Ottawa, Ont.: Lee Valley Tools, Ltd., 1982.

Plumier, Charles. *L'art de tourner*. Lyon: Jean Certe, 1701.

Pollack, Emil, and Martyl Pollack. *A Guide to American Wooden Planes and Their Makers*. Morristown, N.J.: Astragal Press, 1987.

Pye, David. *The Nature and Art of Workmanship.* Cambridge: Cambridge University Press, 1968.

Roberts, Ken. *Wooden Planes in 19th Century America.* 3 vols. Fitzwilliam, N.H.: Ken Roberts Publishing Co., 1983.

Rose, Walter. *The Village Carpenter.* Cambridge: Cambridge University Press, 1938.

Roubo, André-Jacob. *L'art du menuisier.* Paris, 1769–75.

Salaman, R. A. *Dictionary of Tools Used in the Woodworking and Allied Trades, c. 1700–1970.* New York: Charles Scribner's Sons, 1975.

Salivet, Louis-Georges-Isaac. *Manuel du tourneur.* Paris: Bergeron, 1816.

Sellens, Alvin. *Woodworking Planes.* Alvin Sellens, 1978.

Smith, Joseph. *Explanation or Key to the Various Manufactories of Sheffield.* Sheffield, Eng., 1816.

Spon, E., and F. N. Spon. *Spon's Mechanics' Own Book.* New York: E. & F. N. Spon, 1885.

Stokes, Gordon. *Modern Wood Turning.* New York: Sterling Publishing Co., 1979.

Sturt, George. *The Wheelwright's Shop.* Cambridge: Cambridge University Press, 1923.

Tangerman, E. J. *Whittling and Woodcarving.* New York: Dover, 1936.

Taylor, Jay L. B. *Handbook for Rangers and Woodsmen.* New York: John Wiley & Sons, 1916.

Thoreau, Henry David. *Walden; or, Life in the Woods.* New York: Modern Library, 1950.

U.S. Army Corps of Engineers. *Technical Manual 5-225, Rigging and Engineer Hand Tools.* Washington, D.C.: War Department, 1942.

Viires, A. *Woodworking in Estonia.* Springfield, Va.: National Technical Information Service, 1969.

Weygers, Alexander G. *The Modern Blacksmith.* New York: Van Nostrand Reinhold Co., 1974.

White, Lynn. *Medieval Technology and Social Change.* Oxford: Oxford University Press, 1978.

The Young Mechanic. New York: G. P. Putnam & Sons, 1872.

Acknowledgments

Ideas gleaned from conversations with colleagues appear throughout this text. Among the contributors are Curtis Buchanan, Don Carpentier, Brian Coe, Michael Dunbar, Mike Easley, Patrick Edwards, Peter Follansbee, John Reed Fox, Nora Hall, Marcus Hansen, Jeff Headley, Mack Headley, Frank Klausz, Thomas Lamond, Steve Latta, Angus Murdoch, Jim Parker, Jane Rees, Peter Ross, David Salisbury, Ken Schwarz, Robert Self, Keith Thomas, Don Weber, Lyle Wheeler, and Edward Wright. It's entirely my fault if I got anything wrong.

The fastest way to make friends disappear is to ask them to review your manuscript. Thanks to my readers and contributors, Carey Bagdassarian, Ted Boscana, Mack Headley, Kåre Loftheim, Bill Pavlak, and Garland Wood. I owe a special debt to Jay Gaynor, whose insightful and thorough reading and questioning forced out many embarrassing errors.

Who says Underhills can't do anything together? Sister Barbara Ann advised on the graphics, Rachell Underhill undertook the pixel polishing, Jane Underhill assisted on photos, and Eleanor Underhill drew the illustrations.

The artisans shown in the illustrations were surely tempted at times to use their tools as weapons. Thanks to Billy Alexander, Mark Berninghousen, Jacquie Fehon, Frank Grimsley, Mack Headley, Bryant Holsenbeck, Wayne Randolph, Russell Steele, Robert Watson, Bill Weldon, Dan Whitten, and Garland Wood.

The television series *The Woodwright's Shop* is produced by UNC-TV, with generous underwriting support from State Farm Insurance. The Woodwright book series is published by the UNC Press. I work most immediately with my director, Geary Morton, and my editor, David Perry, but I greatly appreciate the contributions of the many talented people at these fine organizations.

Index